AF411463

Urbanization under a Changing Climate

Urbanization under a Changing Climate—Impacts on Urban Hydrology

Editors

Jianxun He
Caterina Valeo
Kasiapillai S. Kasiviswanathan

MDPI • Basel • Beijing • Wuhan • Barcelona • Belgrade • Manchester • Tokyo • Cluj • Tianjin

Editors
Jianxun He
University of Calgary
Canada

Caterina Valeo
University of Victoria
Canada

Kasiapillai S. Kasiviswanathan
Indian Institute of Technology
India

Editorial Office
MDPI
St. Alban-Anlage 66
4052 Basel, Switzerland

This is a reprint of articles from the Special Issue published online in the open access journal *Water* (ISSN 2073-4441) (available at: https://www.mdpi.com/journal/water/special_issues/ Urbanization_Climate_Urban_Hydrology).

For citation purposes, cite each article independently as indicated on the article page online and as indicated below:

LastName, A.A.; LastName, B.B.; LastName, C.C. Article Title. *Journal Name* **Year**, *Volume Number*, Page Range.

ISBN 978-3-0365-0810-8 (Hbk)
ISBN 978-3-0365-0811-5 (PDF)

Cover image courtesy of Jianxun He.

Contents

About the Editors

Jianxun He is an Associate Professor in the Department of Civil Engineering, Schulich School of Engineering at the University of Calgary. Her research interests address several areas of water resources engineering, including low impact development for stormwater management, climate change impact on water resources, and hydrological statistics and extremes.

Caterina Valeo, Ph.D., P.Eng., is a Professor in the Department of Mechanical Engineering at the University of Victoria and an Adjunct Professor in the Department of Civil Engineering, Schulich School of Engineering at the University of Calgary. She specializes in environmental informatics, a branch of geomatics engineering that deals with data creation, analysis, dissemination, and uncertainty, advancing and applying environmental informatics tools for research in low-impact development practices for over 20 years.

Kasiapillai S. Kasiviswanathan is an Assistant Professor in the Department of Water Resources Development & Management at the Indian Institute of Technology Roorkee. His research interests include hydrological modeling, reservoir operation, flood and drought management, optimization, and uncertainty quantification.

Editorial

Urbanization under a Changing Climate–Impacts on Hydrology

Caterina Valeo [1,*], Jianxun He [2] and Kasiapillai S. Kasiviswanathan [3]

[1] Mechanical Engineering Department, University of Victoria, Victoria, BC V8P 5C2, Canada
[2] Department of Civil Engineering, Schulich School of Engineering, University of Calgary, Calgary, AB T2N 1N4, Canada; jianhe@ucalgary.ca
[3] Department of Water Resources Development and Management, Indian Institute of Technology Roorkee, Roorkee 247667, India; k.kasiviswanathan@wr.iitr.ac.in
* Correspondence: valeo@uvic.ca; Tel.: +1-250-721-8623

Citation: Valeo, C.; He, J.; Kasiviswanathan, K.S. Urbanization under a Changing Climate–Impacts on Hydrology. *Water* **2021**, *13*, 393. https://doi.org/10.3390/w13040393

Received: 28 January 2021
Accepted: 29 January 2021
Published: 3 February 2021

Publisher's Note: MDPI stays neutral with regard to jurisdictional claims in published maps and institutional affiliations.

On a global scale, urbanization and climate change are two powerful forces that are reshaping ecosystems and their inhabitants. The scale and manner in which these two phenomena are changing the globe may differ, but any orchestrated response to one cannot be very effective without due consideration to the other. For all intents and purposes, anthropogenic response and adaptation are localized, disconnected, and single-purposed for various reasons including economic, social, political, and scientific. Thus, there remains a lack of understanding of how the holistic "sum" of these two forces can be assessed. This is not surprising, nor a criticism, but a simple fact that is difficult to manage. Researching the impacts of these two phenomena individually will likely only provide short-term results that are only marginally effective.

In many respects, urban hydrology is often framed in terms of increases in imperviousness, increases in peak flowrates, volumes and risk of flooding, increases in contamination, reduction to groundwater recharge, management via the end-of-pipe infrastructure, etc. Climate change impacts are often viewed in terms of changes to fundamental urban hydrological input parameters: changes to precipitation intensity, peaks, total volumes, frequency, duration [1], inter-event period length, minimum and maximum temperatures, and therefore, changes to response, planning and disaster mitigation (flooding or droughts). Stormwater management (i.e., our response to urban hydrology), has evolved over the last several decades to observe the need for sustainability–an often overly used term to convey anything "environmentally-conscious"; but here it is used to recognize that the needs of future generations are pivotal in its definition; and therefore, planning for a future that incorporates climate change is an integral part of all sustainable stormwater management. In urban hydrology, a technology that has been embraced in the name of sustainability is the Low Impact Development technology or LID. LID technology is designed to bring the post-development hydrograph back to its pre-developed state as much as possible. Thus, treatment involves reducing peak flowrates, prolonging timing and in many of these technologies, reducing and potentially eliminating contaminants arising in urban areas. Furthermore, urban water management aims to improve system resiliency and sustainability given that the principle of stationarity may be invalid under urbanization and climate change. An increasing role for LID in future planning that mitigates the effects of urbanization *and* climate change by impacting the fundamental hydrological parameters noted above, is apparent in this special issue of *Water*.

While at first glance the submissions to this special issue may seem disparate, they are however, all intimately connected to urban hydrology and most importantly, they span the wide range of scales of observation and consequences of urban hydrology. From the near microscopic scale [2] to meter scale [3,4] to km scale [5–8], this special issue effectively brings awareness and attention to the range of scales at which these phenomena manifest, and the responses that are necessary if societies are to truly become resilient. Spatial scale is not the only variant—temporal scales also vary across the study of urbanization and climate

1

change impacts. Although often receiving less attention in observation and modelling than does spatial scaling, temporal scaling is no less important in truly sustainable responses.

In Zhao and Zhu [2], the authors provide a highly focused look at the leaching behaviours of zinc, lead, copper, and chromium; four highly problematic contaminating metals in urban runoff that are expected to increase in the future. The examination is focused on permeable pavements—a very popular and nearly pervasive LID in many parts of the globe. The authors found that while several fine-scale chemical, mechanical and physical processes within the pavement gave rise to their observations, they observed that in several instances, leaching concentrations changed with contact time (depending on the metal in question). This suggests that depending on the permeable pavement design, longer inter-event periods arising from climate change may lead to less incidents of flushing, and may in turn, lead to variability in metal contamination in stormwater at the end of the system. This can have adverse consequences for future planning and management.

At a slightly larger scale, outside the laboratory, but still focusing on LID technologies, Li et al. [4] examined the LID technology of bioretention cells for the curbside treatment of stormwater quantity and quality. Specifically looking at roadside bio-retention (RBR) facilities, the authors wanted to assess the influence of several fundamental assumptions involved in the design, and in particular, the importance of inlet hydraulics and how the spatial distribution of inflow along the RBR influence overall runoff control performance—something the authors note is rarely considered in LID modeling. The scale of focus here is from cm to meters, and this study included a very well written and concise literature on how RBR's function, and the corresponding research in the lab, field and mathematical modelling. The important insight gained in their research include the fact that the effective length of an RBR, which depends upon the inflow rate as well as the perforation arrangement of the inflow distribution pipe, has an effect on the overall runoff control performance. In fact, the authors found that no matter how well an RBR is designed, the inlet dictated overall runoff control performance. Again, this can have serious consequences for future planning in stormwater management.

At a larger scale, but still focusing on LID technologies, Mahmoodzadeh et al. [3] examined green roofs as a means for affecting the energy budget of school buildings across a variety of North American climates and cities. Using a popular energy modelling software, the authors evaluated the relevance of green roof vegetation characteristics (such as leaf albedo, for example) on the roof's ability to maintain or improve the insulative value of the school building. This was done for four different climate areas spanning a range of temperatures and humidity levels. The authors found that even with the simplifying assumptions arising from the model, in climates where vegetation is active (outside of winter seasons where vegetation is in senescence), leaf area index had a major influence on cooling load reduction in all four cities. While optimal vegetation characteristics could be determined with the model, they were certainly climate dependent. The study highlighted the fact that increasing vegetation biomass in urban areas can help to reduce aspects of the heat island effect, and the role of vegetation in LID may become more pronounced as climate change impacts amplify.

Expanding the spatial scale even further, two articles ([5,7]) looked at locating LIDs in urban areas by creating a new approach to identify optimal locations for LIDs. Kaykhosravi et al. [7] presented a highly novel approach on how to incorporate LID in planning through a useable, simple, but rigorous index for determining the optimal location of an LID. The authors proposed a geospatial, physically-based framework integrated with a multi-criteria decision-making model that considered both environmental and socioeconomic factors. The LID Demand Index, or LIDDI, was able to effectively determine the optimal locations for LID demand across a large Canadian city (Toronto, Canada) to help reduce the risk of flooding and at the same time, provide the socioeconomic and environmental benefits of LID. This incredible new tool allows decision-makers to conduct future land-use planning quickly and holistically, in order to make landscape and infrastructure planning under a changing climate more resilient and effective.

At the municipal city scale, Kaykhosravi et al. [5] provided an examination of how climate change and urbanization combine to increase the risk of flooding in urban areas and how the demand for the stormwater mitigating benefits of LIDs will improve or change under these changing conditions. This seminal work shows conclusively using a variety of scenarios for both urbanization and climate change, that demand for LIDs will increase and how; and most importantly, shows how the demand varies depending on the climate, and of course, the nature of the increase in urbanization and urban form.

Further increasing the spatial scale is an examination of the impacts of climate and land-use on bacterial concentrations in stormwater. Xu, et al. [6] looked at the relationship between observations of fecal coliform concentrations in urban stormwater runoff from the lower Vancouver Island region in Canada. The work demonstrated a clear relationship historically between increases in fecal coliform (FC) concentrations and minimum temperatures (positive), antecedent dry period (positive), and higher urban area proportions (positive,) and with negative relationships to precipitation volumes. Using data driven modelling techniques, the study demonstrated that increasing urbanization leads to increases in FC but that conversely to other studies, longer periods of drier weather will likely lead to an increase in bacterial loadings from urban watersheds. Longer, drier periods with increasing daily temperature minimums may lead to higher growth in FC levels. Precipitation that normally dilutes and washes-off bacteria loadings and tends to reduce FC observations in stormwater may be altered. If future climate change scenarios suggest a combination of longer, drier inter-event periods with changes to precipitation volumes at the same time, urban regions may see increases in FC levels in stormwater in the future.

At the largest spatial scale in this special issue of *Water*, Mohanavelu et al. [8] used statistics on a large database of point-scale well observations to examine the influence of both climate change and urbanization on groundwater levels from several cities in India. Recognizing that large aquifers are the predominant water source for many urban areas in India, the authors undertook to determine the signal in dropping groundwater levels arising from both climate and the rapid urbanization being experienced across the country. The authors found that non-stationarity was observed across seasons suggesting water levels are related to land use changes (in turn related to increasing urbanization) and, potentially, climate variability. The authors carefully noted that historical, irreversible aquifer depletion caused by a high rate of pumping over a long time could have rendered the effects of climate variability and land use changes as negligible. Less interdependence between seasonality (based on monsoon) and groundwater levels was observed, which might have been caused by the impact of climate variability on groundwater recharge. The research suggests that rainfall–groundwater level relationships have been seriously affected across all the studied cities and possibly due to reductions in monsoon rainfall, which may have resulted from climate change.

Overall, this special issue of *Water* was very successful in highlighting the need for incorporating both urbanization and climate change in tandem in future planning, future research, and problem-solving, at all the relevant spatial and temporal scales.

Author Contributions: C.V. drafted the initial version, J.H. and K.S.K. revised and edited the final version. All authors have read and agreed to the published version of the manuscript.

Funding: This research received no external funding.

Acknowledgments: The authors would like to thank all the contributors who submitted their research articles to this special issue and, in particular, a very special thank you to all the reviewers who reviewed the submissions and provided their insightful input. They greatly enhanced the quality of the submissions to make this a great special issue in *Water*.

Conflicts of Interest: The authors declare no conflict of interest.

References

1. He, J.; Valeo, C.; Bouchart, F.J.C. Enhancing urban infrastructure investment planning practice for a changing climate. *Water Sci. Technol.* **2006**, *53*, 13–20. [CrossRef] [PubMed]
2. Zhao, Y.; Zhu, Y.-T. Metals Leaching in Permeable Asphalt Pavement with Municipal Solid Waste Ash Aggregate. *Water* **2019**, *11*, 2186. [CrossRef]
3. Mahmoodzadeh, M.; Mukhopadhyaya, P.; Valeo, C. Effects of Extensive Green Roofs on Energy Performance of School Buildings in Four North American Climates. *Water* **2020**, *12*, 6. [CrossRef]
4. Li, J.; Alinaghian, S.; Joksimovic, D.; Chen, L. An Integrated Hydraulic and Hydrologic Modeling Approach for Roadside Bio-Retention Facilities. *Water* **2020**, *12*, 1248. [CrossRef]
5. Kaykhosravi, S.; Abogadil, S.; Khan, U.T.; Jadidi, M.A. The Low-Impact Development Demand Index: A New Approach to Identifying Locations for LID. *Water* **2019**, *11*, 2341. [CrossRef]
6. Xu, K.; Valeo, C.; He, J.; Xu, Z. Climate and Land Use Influences on Bacteria Levels in Stormwater. *Water* **2019**, *11*, 2451. [CrossRef]
7. Kaykhosravi, S.; Khan, U.T.; Jadidi, M.A. The Effect of Climate Change and Urbanization on the Demand for Low Impact Development for Three Canadian Cities. *Water* **2020**, *12*, 1280. [CrossRef]
8. Mohanavelu, A.; Kasiviswanathan, K.S.; Mohanasundaram, S.; Ilampooranan, I.; He, J.; Pingale, S.M.; Soundharajan, B.-S.; Diwan Mohaideen, M.M. Trends and Non-Stationarity in Groundwater Level Changes in Rapidly Developing Indian Cities. *Water* **2020**, *12*, 3209. [CrossRef]

Article

Metals Leaching in Permeable Asphalt Pavement with Municipal Solid Waste Ash Aggregate

Yao Zhao * and Ya-Ting Zhu

College of Civil Engineering, Nanjing Forestry University, 159 Longpan Road, Nanjing 210037, China; zytnjfu@163.com

* Correspondence: zhaoyao@njfu.edu.cn; Tel.: +86-25-854-277-58

Received: 18 September 2019; Accepted: 16 October 2019; Published: 21 October 2019

Abstract: The leaching behaviors of four heavy metals (Zn, Pb, Cu and Cr) from unbounded municipal solid waste incineration-bottom ash aggregate (MSWI-BAA) and permeable asphalt (PA) mixture containing MSWI-BAA were investigated in the laboratory. The horizontal vibration extraction procedure (HVEP) test and a simulated leaching experiment were conducted on MSWI-BAA with three particle sizes, but only the simulated leaching experiment was carried out on a type of PA specimen (PAC-13) with and without these MSWI-BAAs. Leaching data were analyzed to investigate the leaching characteristics, identify the factors affecting leaching and assess the impact on the surrounding environment. Results indicated that the leaching process was comprehensively influenced by contact time, leaching metal species and MSWI-BAA particle size, regardless of MSWI-BAA alone or used in PAC-13 mixture. The leaching concentrations of Cr, Zn and Pb from MSWI-BAA in HVEP testing was strongly related to MSWI-BAA particle size. The use of MSWI-BAA in PAC-13 mixture did not change the basic tendency of heavy metal leaching, but it led to an increase of Cr and Zn in leachate overall. The leachate from the MSWI-BAA and PAC-13 mixture with MSWI-BAA was shown to be safe for irrigation and would have very little negative impact on surrounding surface and underground water quality.

Keywords: permeable asphalt; heavy metal; leaching behavior; MSWI-BAA; stormwater

---

1. Introduction

Incineration with energy recovery is recommended as a preferred option for dealing with municipal solid waste (MSW) to effectively reduce the original waste volume and mass by approximately 90% and 70%, respectively, and generate electricity and heat [1–3]. However, a considerable amount of residual material is still generated after the incineration process: typically MSW incineration fly ash (MSWI-FA) and MSW incineration bottom ash (MSWI-BA) [4,5]. MSWI-BA accounts for nearly 80% of the total residual by mass and is complex, consisting of combustion residue and non-combustible constituents of the waste feed [6–8]. In Europe and Asia, MSWI-BA is often classified as non-hazardous waste [9]. For this reason, MSWI-BA is commonly discarded in landfills.

In 2017, 84.63 million tons of MSW were treated by incineration in China, accounting for nearly 39% of the total mass [10], which produced approximately 20.31 million tons of MSWI-BA. It is reported that the amount of MSWI-BA generated is expected to exceed 28 million tons by 2020 according to the National Thirteenth Five-Year Plan of China [11,12]. If all these MSWI-BAs are to be disposed of in landfill sites; around 12.73 million m^3 of landfill space (calculated by the density of 2.20 g/cm^3 [13,14]) will be needed. Therefore, MSWI-BA treatment has been a tremendous challenge to most Chinese cities with high population densities and limited land resources.

Traditionally, landfilling is regarded as the most convenient and inexpensive approach for disposal of MSWI-BA. However, it results in significant environmental problems. Studies indicate that MSWI-BA contains various heavy metals such as zinc (Zn), lead (Pb), copper (Cu) and chromium (Cr), which are present in high concentrations [3,15,16]. The leaching of these metals from MSWI-BA when exposed to rainwater can seriously contaminate the surrounding sensitive recipients, including soil, surface and sub-surface water bodies [5,17,18]. Environmental safety, thus, has become a great concern for MSWI-BA management.

The recycling of MSWI-BA as road construction material, especially considering the reduction of natural aggregate usage, is recommended as an important management option and has gained worldwide attention. Numerous studies were conducted on the physicochemical and engineering properties of MSWI-BA. It is concluded that MSWI-BAA made by fresh MSWI-BA after being pretreated, has a good particle size distribution and similar properties to natural aggregate and is suitable to be used in asphalt mixtures and cement concrete [4,5,9,13,14,19,20]. Some studies have consequently used MSWI-BAA to partially replace natural aggregate in asphalt mixtures. It is reported that the substitution of MSWI-BAA with coarse and fine natural aggregates in traditional, dense hot-mix asphalt (HMA) can meet the technical requirements, but the characteristics of being lightweight, and having a smaller specific weight, compared to natural aggregate, must be seriously considered [14,21,22]. It is, therefore, suggested that MSWI-BAA is more suitable for use in low-traffic-volume roads [23].

Rapid urbanization coupled with climate change is placing increasing pressures on urban stormwater management [24]. The increase in impervious surfaces with asphalted roads and rooftops significantly increases stormwater volumes and peak flowrate, while also decreasing stormwater quality and impeding groundwater recharge [25–27]. Permeable asphalt (PA) pavement is a popular and practical, low-impact development (LID) technology that serves as an ideal alternative to conventional low-traffic-volume pavement because it can help to address the issues stated above by providing in situ restoration of the urban hydrological cycle and reducing the need for traditional stormwater facilities [28]. PA pavement consists of various layers with porous materials, which not only allow stormwater to infiltrate into the ground, unlike conventional asphalt pavement, but can also simultaneously remove pollutants (e.g., total suspended solids (TSS) and heavy metals) from stormwater runoff on site, thus raising their value as a LID option in current urban development [25,29–31]. As the PA mixture (the surface material of the PA pavement) requires relatively low strength compared to conventional dense asphalt mixtures, researchers have explored the feasibility of recycling MSWI-BA aggregate (MSWI-BAA) in PA mix designs. The results are encouraging, indicating that PA mixtures containing MSWI-BAA have better performance than those without, and the replacement ratio can be up to 80% [23].

On the other hand, the potential environmental impacts associated with the use of MSWI-BAA in asphalt mixtures are of great concern [21,32–34]. Results indicate that the leaching concentrations of heavy metals (e.g., Pb, Zn, Cu, Cr and Cd) from HMAs containing MSWI-BAA are significantly reduced compared to unbounded MSWI-BAA. It has been concluded that there is very little environmental risk for substituting MSWI-BAA for natural aggregate in dense asphalt mixtures. Previous investigations have provided valuable information for the leaching characteristics of heavy metals from dense asphalt mixtures containing MSWI-BAA; however, there are few studies focusing on PA mixtures containing MSWI-BAA. High voids in PA mixtures provide benefits for stormwater management in urban areas, but it can also lead to an increase in contact between the PA material and the infiltrated stormwater during wet weather.

To encourage the utilization of MSWI-BAA in PA mixtures, investigating the leaching behavior of unbounded MSWI-BAA and within PA mixtures to identify environmental consequences is warranted. The objectives of this study are to investigate and compare the heavy metal leaching of unbounded MSWI-BAA and PA mixtures containing both the coarse and fine MSWI-BAAs at optimal replacement ratios. In this study, heavy metals present in MSWI-BAAs with three particle sizes (0.075–2.36 mm, 2.36–4.75 mm and 4.75–9.5 mm) were identified based on X-ray Fluorescence (XRF) tests. The leaching behaviors of heavy metals in MSWI-BAA with three particle sizes were investigated with HVEP tests and simulation experiments. The leaching behaviors of PA mixtures containing these MSWI-BAAs were examined by simulation experiments. The difference in leaching behavior between unbounded MSWI-BAA and PA with MSWI-BAA were compared and analyzed. The leaching data were also compared to limit values in Chinese standards for surface water, subsurface water and irrigation water, to assess the environmental risk for the utilization of MSWI-BAA in PA mixtures.

2. Materials and Methods

2.1. Materials

2.1.1. MSWI-BAA

The MSWI-BAA used in this study was made by the MSWI-BA collected from a waste-to-energy facility located in Nanjing City, China. After the incineration process, fresh MSWI-BA was water-quenched immediately and air-dried for 7 days, and then delivered to the lab. The received MSWI-BA was naturally weathered in the lab for another 90 days for potential hydration and stabilization prior to pretreatment. The processes used for MSWI-BA pretreatment included impurity removal, magnetic separation, crushing and sieving. Three MSWI-BAA sizes were prepared: 0.075–2.36 mm, 2.36–4.75 mm and 4.75–9.5 mm. It should be noted that both sampling and preparation were carried out in accordance with Chinese standards of HJ/T 20-1998, HJ/T 298-2007 and JTG E42-2005 [35–37].

To identify the main species and contents of heavy metals in MSWI-BAAs of different sizes, XRF was utilized. Three groups of MSWI-BAA samples were oven dried at 105 °C until a constant mass was obtained, and then the samples were milled into powder (<0.075 mm). After that, these powder samples were separately packed in sealable polypropylene bags prior to further testing. The weight percentages of all the elements above detection limit are shown in Table 1. It can be noted that Zn, Cu, Cr and Pb are present in relatively high levels in these three MSWI-BAA samples, accounting for approximately 0.7–2.2% of the total mass. Thus, these four metals were selected as the tracer metals for the leaching tests in this study.

Table 1. XRF analysis results of MSWI-BAA samples.

Element	MSWI-BAA (% by Weight)		
	0.075–2.36 mm	2.36–4.75 mm	4.75–9.5 mm
Ca	51.55	44.56	43.97
Si	12.85	17.27	19.60
Al	6.94	8.65	8.82
Cl	4.78	4.88	5.67
Fe	5.92	5.65	4.77
S	4.49	3.49	3.61
Mg	2.26	2.44	2.61
P	2.66	3.90	2.23
Ti	1.98	1.65	1.63
K	2.16	2.46	2.53
Na	1.59	2.14	3.22
Zn	1.32	1.92	0.48
Cu	0.22	0.12	0.10
Cr	0.17	0.13	0.10
Pb	0.19	0.03	0.03
Sr	0.12	0.08	0.04
Ba	0.60	0.41	0.44
Total	99.80	99.78	99.85

2.1.2. PAC-13 Mixture Containing MSWI-BAA

A type of PA mixture with a maximum aggregate size of 13.2 mm, PAC-13 mixture, was designed in accordance with the Technical Specification for Permeable Asphalt Pavement (CJJ/T 190-2012) [38].

To focus on the realistic leaching behavior of PAC-13 containing MSWI-BAA on site, PAC-13 mixtures used in this study were designed with the same aggregate graduation and air void content (20%), as shown in Figure 1. Three sizes of MSWI-BAA (0.075–2.36 mm, 2.36–4.75 mm and 4.75–9.5 mm) were separately used as partial replacements of the same aggregate size in the PAC-13 mixture with optimal replacement ratios of 30%, 20% and 30% (by aggregate weight). Thus, four types of PAC-13 specimens, including three PAC-13 specimens containing MSWI-BA and one specimen without (control), were used in this study. The optimal replacement ratio of MSWI-BAA for each mixture was determined to satisfy the technical requirements for PAC-13 mixtures.

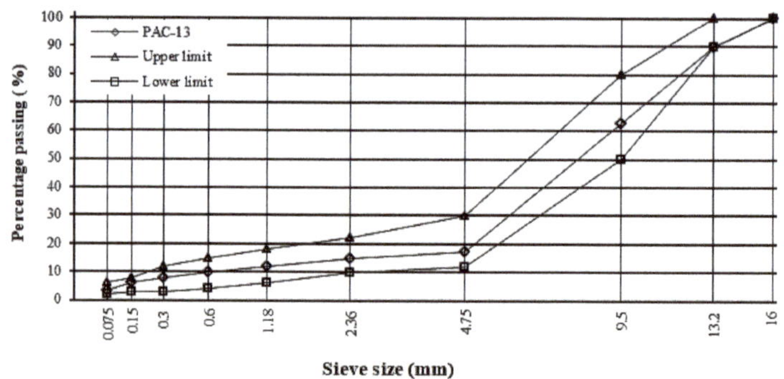

Figure 1. Aggregate gradation of PAC-13 mixture.

All the PAC-13 specimens were compacted by a Marshall hammer with 50 blows per side at 170 °C, according to Method T0702 [39]. Each specimen was prepared in a 101.6 mm diameter mold with height 63.5 ± 1.3 mm. The optimum asphalt contents of 5.4%, 5.1%, 5.7% and 4.9% were separately used for three PAC-13 specimens containing MSWI-BAA with particle sizes of 0.075–2.36 mm, 2.36–4.75 mm and 4.75–9.5 mm, and one specimen without MSWI-BAA. It should be noted that all the materials utilized for preparing these specimens satisfied the technical requirements of CJJ/T 190-2012 [38].

2.2. Methods

2.2.1. Leaching Test of MSWI-BAA

To assess the environmental risk resulting from the possible release of heavy metals (Zn, Pb, Cu and Cr) from MSWI-BAA in unbounded form, the standard test and a simulated leaching experiment were conducted in the lab.

The HVEP test (as shown in Figure 2) was conducted in accordance with the Chinese standard (HJ 557-2009) [40] to evaluate the potential maximum leaching toxicity of heavy metals from unbounded MSWI-BAA when exposed to stormwater. For preparing the liquid waste sample, 100 g of MSWI-BAA sample was set with 1 L distilled water in a flask after passing through a 3 mm sieve. The mixture sample was vibrated with frequency of 110 times/min and amplitude of 40 mm for 8 h, and then rested at room temperature (20 °C) for 16 h. The liquid extracted was then separated from the solid phase by filtration through a 0.45 μm glass fiber filter and then stored in the refrigerator at 4 °C before digestion. A total of three MSWI-BAA samples and one blank were tested.

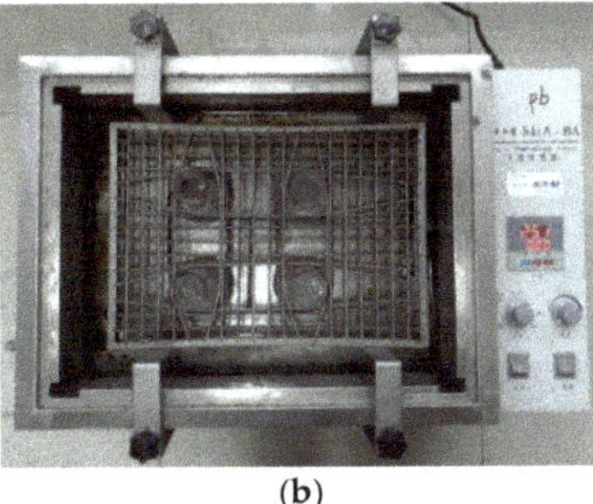

(a)   (b)

Figure 2. HVEP test in this study: (**a**) Liquid waste samples; (**b**) the horizontal-vibration machine.

The simulated leaching experiment was designed to assess the realistic leaching of heavy metals from unbounded MSWI-BAA exposed to stormwater over a long period of time. The simulated experiment was continued for 28 days. Although the pH level is suggested as one of the main factors influencing MSWI-BA's leaching behavior [17], distilled water with an initial pH value of 6.8 was used here to simulate the stormwater. The liquid/solid (L/S) ratio of 10:1 was chosen based on results from [41] and the HVEP test method. For each experimental sample, 100 g of MSWI-BAA was mixed with 1 L of synthetic stormwater in a capped flask. Timing commenced immediately after application of the synthetic stormwater. At each sampling event, 10 mL of leachate was collected from a flask using a pipette, dispensed into a polypropylene sample bottle, and then 3 mL of nitric acid was added to the sample to keep it stable. Samples were stored in the refrigerator at 4 °C until further analyses. It should be noted that the first leachate sample was collected immediately after the application of synthetic stormwater to the MSWI-BAA sample. Herein, a total of three MSWI-BAA samples with different sizes were used and 14 sampling events were conducted during the experiment for each sample.

2.2.2. Simulated Leaching Experiment of PAC-13 Mixture Containing MSWI-BAA

The experimental procedure for the PAC-13 mixtures containing MSWI-BAA was the same as that for the unbounded MSWI-BAA. Four PA specimens were used, as shown in Figure 3: (1) PAC-13 with 30% MSWI-BAA of 0.075–2.36 mm; (2) PAC-13 with 20% MSWI-BAA of 2.36–4.75 mm; (3) PAC-13 with 30% MSWI-BAA of 4.75–9.5 mm; and (4) PAC-13 without MSWI-BAA. For each PA specimen, a total of 14 sampling events were conducted during the experiment.

Figure 3. PAC-13 specimens with and without a municipal solid waste incineration-bottom ash aggregate (MSWI-BAA) used in the simulated leaching experiment.

2.2.3. Sample Determination and Data Analysis

After the leaching tests, all of the leachate samples were first digested by a microwave digestion procedure. The total concentrations of Zn, Cu, Cr and Pb in leachates were then determined by inductively coupled plasma-mass spectrometry (ICP-MS).

For data analysis, Microsoft Excel® and SPSS® were used to conduct a variety of tests on the data to identify the difference of leaching characteristics and determine the factors affecting heavy metal leaching. The two-factor analysis of variance (ANOVA) was conducted on the data from the HVEP test while the one-factor ANOVA was conducted on the data from simulated experiments. Linear regression was also used to determine if linear relationships were likely.

3. Results

3.1. HVEP Results

Table 2 presents the leaching concentrations of heavy metals Zn, Cu, Cr and Pb from unbounded MSWI-BAA samples with three particle sizes obtained through the HVEP test. For all MSWI-BAA samples, the leaching concentration of Cr was the highest, followed by Cu, Zn and Pb. The leaching concentration of Cr from MSWI-BAA with particle size of 4.75–9.5 mm was the highest, while Pb leaching from MSWI-BAA of the same particle size was the lowest. For each heavy metal, the leaching concentration seemed to be related to MSWI-BAA particle size. Moreover, an MSWI-BAA with a larger particle size was likely to release more Cr and Zn, and an MSWI-BAA with smaller particles was likely to release more Pb. These results suggest that the heavy metal leaching from unbounded MSWI-BAA in the HVEP test is a result of the combined effect of both the leaching metal species and MSWI-BAA particle size.

Table 2. Leaching concentrations of heavy metals from MSWI-BAA by the HVEP test.

Heavy Metal [1]	Leaching Concentration (µg/L)			Limit Value (µg/L)		
	0.075–2.36 mm	2.36–4.75 mm	4.75–9.5 mm	GB/T 14848 (Class III)	GB 3838 (Class II)	GB 5084
Cr	45.64	50.73	106.08	50 (VI) [2]	50 (VI) [2]	100 (VI) [2]
Cu	28.93	20.49	26.88	1000	1000	1000
Zn	4.77	7.04	6.79	1000	1000	2000
Pb	5.35	0.32	0.24	10	10	100

[1] Heavy metal concentration in leachate sample is the total concentration; [2] these limit values are intended for Cr(VI).

3.2. Simulated Leaching Experiment Results

3.2.1. Leaching Process of MSWI-BAA

Figures 4 and 5 present the results of the simulated leaching experiment using unbounded MSWI-BAA with three different particles. As demonstrated in Figure 4, overall, the leaching concentrations of heavy metals changed over contact time with varying tendencies. Cr and Cu leaching values showed a fluctuating, increasing trend with contact time, while the others, surprisingly, did not. During the first 10 days, Zn and Pb showed a fluctuating, decreasing trend, and then remained at a relatively low level. In addition, Cr concentration in leachate increased faster than Cu throughout the experiment, and it remained the highest level in the second half of the period. This result indicates that the leaching trend of heavy metal from unbounded MSWI-BAA is mainly dependent on the type of leaching metal rather than the MSWI-BAA particle size.

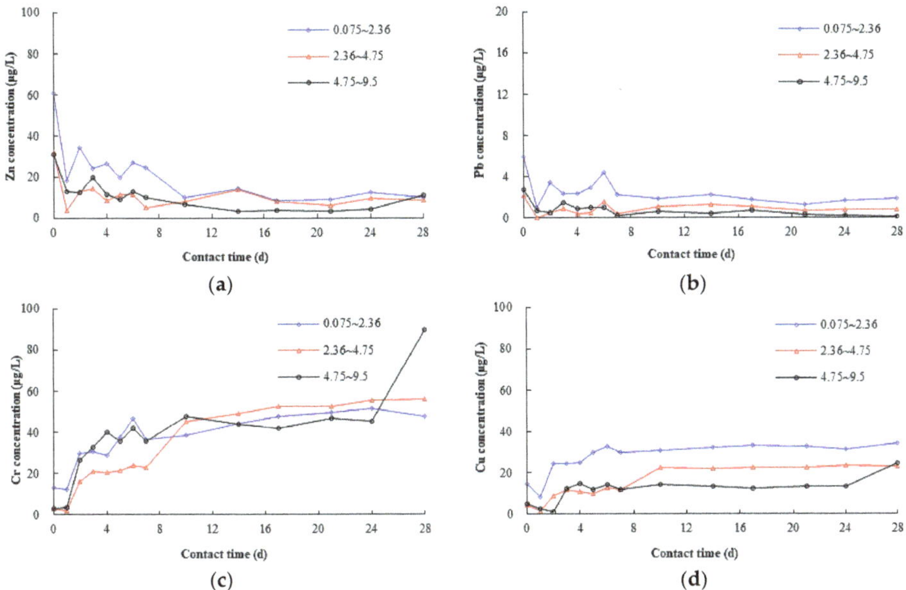

Figure 4. Variation of leaching concentrations of each heavy metal with contact time for MSWI-BAA:
(**a**) Zn; (**b**) Pb; (**c**) Cr; and (**d**) Cu.

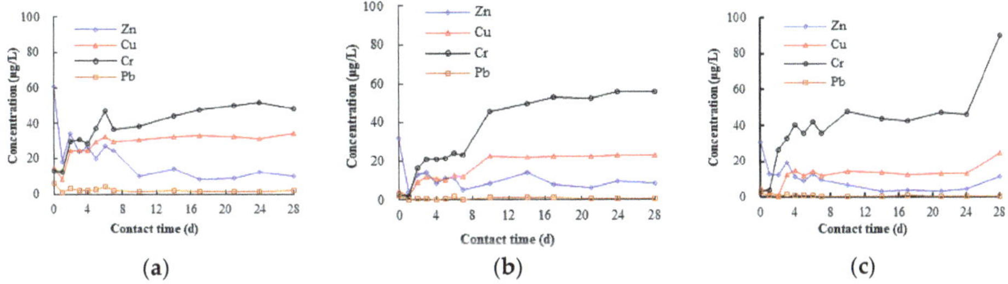

Figure 5. Variation of leaching concentrations of heavy metals with contact time for different MSWI-BAA
particle sizes: (**a**) 0.075–2.36 mm; (**b**) 2.36–4.75 mm; (**c**) 4.75–9.5 mm.

Regarding the leaching level of each heavy metal (as shown in Figure 4), the leaching concentration
of Pb and Cu from MSWI-BAA with the smallest particle size (0.075–2.36 mm) generally remained
at the highest level, followed by the particle size of 2.36–4.75 mm and 4.75–9.5 mm. But Cr showed
a different pattern. Cr concentration in leachate from MSWI-BAA with 2.36–4.75 mm remained at
the lowest level in the first 10 days, but remained at the highest level during the following 17 days.
Moreover, in the experimental period, the highest leaching concentration (90.06 µg/L) was found for
Cr from MSWI-BAA with a particle size of 4.75–9.5 mm at the last sampling event, while the lowest
leaching concentration (0.17 µg/L) was found for Pb from MSWI-BAA of the same particle size at the
last sampling event, which is consistent with the observation in the HVEP test. These results suggest
that the overall level of each heavy metal in leachate seems to be affected by MSWI-BAA particle size,
with the MSWI-BAAs with smaller particles leading to higher levels of leaching of Zn, Pb and Cu.

As shown in Figure 5, for each MSWI-BAA sample, the changes in Zn, Pb, Cr and Cu concentration
in leachate over contact time are similar, but the levels are different. In general, the highest level
measured in this experiment was Cr, followed by Cu, Zn and Pb. For the MSWI-BAA of 4.75–9.5 mm,

a remarkable increase in leaching concentration was observed for Cr, Cu and Zn at the last sampling event, which means their solid and liquid phase equilibriums may not be attained even after 28 days in the leaching period. The results from the simulated leaching experiment validate the suggestion that the leaching of heavy metals from MSWI-BAA is a very slow process, which is comprehensively influenced by contact time, the type of heavy metal leaching and MSWI-BAA particle size, under experimental conditions. With the increasing contact time, the leaching trend is mainly dependent on the type of leaching metal, while the leaching level is greatly affected by the MSWI-BAA particle size.

3.2.2. Leaching Process of PAC-13 with MSWI-BAA

Figure 6 presents the leaching data of each heavy metal varying with contact time, while Figure 7 shows the leaching data of each PA specimen varying with contact time. Figure 6 shows that the leaching concentrations of Zn, Pb, Cu and Cr change in a similar manner with contact time regardless of whether MSWI-BAA is present. Leaching concentrations of Zn, Pb and Cu were stable within a narrow range between 19.14 and 29.15 µg/L. By contrast, the leaching concentration of Cr increased with contact time, and increased faster in the last 7 days. With the exception of some sampling points, Cr and Zn from the control specimen resulted in lower leaching concentrations compared with those specimens containing MSWI-BAA. This might mean that the use of MSWI-BAA in PAC-13 mixtures may lead to increased leaching of Cr and Zn, but these results did not seem to suggest that the use of MSWI-BAA to partially replace a natural aggregate in PAC-13 mixtures can change the basic leaching trend of heavy metals during the experimental period.

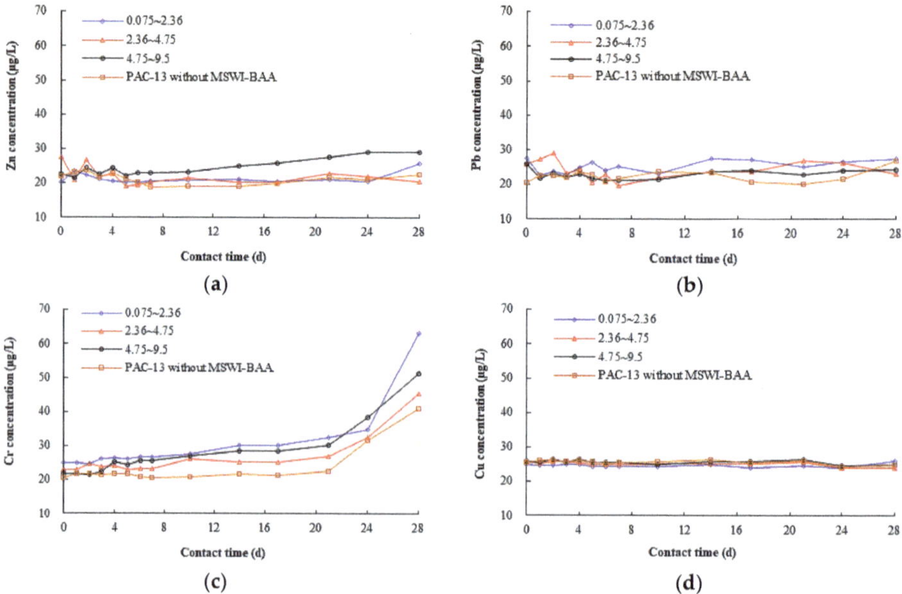

Figure 6. Variation of leaching concentrations of each heavy metal with contact time for PAC-13 specimens with and without MSWI-BAA: (**a**) Zn; (**b**) Pb; (**c**) Cr; and (**d**) Cu.

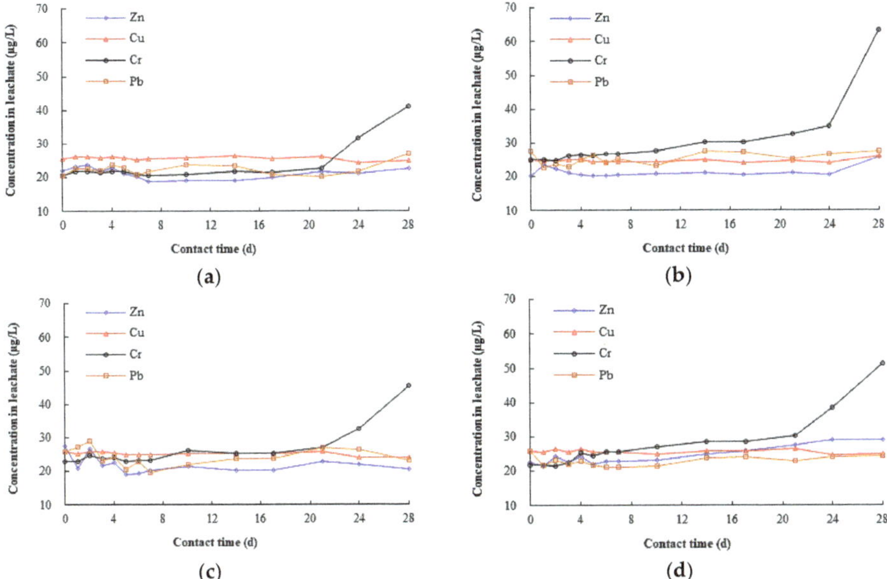

Figure 7. Variation of leaching concentrations of four heavy metals with contact time for PAC-13 specimens with and without MSWI-BAA: (**a**) PAC-13 without MSWI-BAA; (**b**) PAC-13 with MSWI-BAA of 0.075–2.36 mm; (**c**) PAC-13 with MSWI-BAA of 2.36–4.75 mm; and (**d**) PAC-13 with MSWI-BAA of 4.75–9.5 mm.

For the three PAC-13 specimens containing MSWI-BAA, the leaching trend of each heavy metal was similar, regardless of the MSWI-BAA particle size. In general, the leaching of Cu was the most stable throughout the experiment and the concentration fluctuated in a narrow range between 24.09 and 26.39 µg/L. The leaching concentrations of Pb and Zn showed a fluctuation in a relatively wider range. However, the leaching concentration of Cr showed an increasing trend with contact time. These observations indicate that different types of heavy metals produce different leaching trends and leaching levels because some heavy metals are affected by particle size.

Figure 7 shows that for all PAC-13 specimens, the leaching trends of Zn, Cu, Cr and Pb are similar. For the control specimen, Zn, Pb and Cu showed a relatively stable leaching process; however, Cr showed an increase during the last 7 days. Regarding the mean leaching level, Cu was the highest (25.77 µg/L), followed by Cr (23.67 µg/L), Pb (22.52 µg/L) and Zn (21.35 µg/L). For PAC-13 containing MSWI-BAA, the order of mean leaching concentrations of these four heavy metals varied with particle size used, but the mean leaching level of Cr was the highest in all cases. These results suggest that the type of heavy metal is an important factor affecting the leaching trends and levels with or without MSWI-BAA. In addition, the particle size of MSWI-BAA used in the PAC mixture seems to have an effect on the leaching level of only some heavy metals.

4. Discussion

4.1. Leaching of Heavy Metals from MSWI-BAA by HVEP Test

Results showed that Pb and Zn leaching values were much lower than those of Cu and Cr in the HVEP test, which indicates that Cr and Cu are easier to extract compared to Pb and Zn. However, Seniunaite et al. [3] reported relatively higher leaching concentrations for Zn and Pb from MSWI-BA by the standard method of EN 12457-2. Thus, although the experimental conditions (e.g., temperature, L/S ratio, leaching time and leachant) were the same in both standard tests, opposite results were found. This was possibly due to the different heavy metal contents and chemical compositions of the two MSWI-BA samples used.

The HVEP data also showed that the leaching concentration of each heavy metal varied with various MSWI-BAA particle sizes (as shown in Table 2). Two-factor ANOVA was conducted on the leaching data, and the results show a significant difference in leaching concentration among the four heavy metals ($p = 0.011$) and no significant difference in leaching concentration among three MSWI-BAA samples ($p = 0.43$). This result provides evidence for the suggestion that the type of heavy metal is the main factor affecting the heavy metal leaching from MSWI-BAA.

Linear regressions were attempted using the leaching data for each heavy metal. The correlation coefficients between MSWI-BAA particle size and leaching concentrations of Cr, Zn, Pb and Cu are 0.81, 0.66, -0.76 and 0.054, respectively. This means that the leaching of Cr, Zn and Pb are strongly related to MSWI-BAA particle size. For Cr and Zn, the smaller the MSWI-BAA particle size, the higher the leaching concentration will be, while the opposite is true for Pb.

4.2. Leaching of Heavy Metals from MSWI-BAA by Simulated Experiment

As different leaching behaviors were found for heavy metals from unbounded MSWI-BAAs in the simulated leaching experiment, the one-factor ANOVA was conducted on the leaching data. Significant differences in leaching concentration were found among MSWI-BAA samples ($p < 0.01$, for the three groups). Combining the results from the HVEP test, it was demonstrated that the leaching trend and leaching level of heavy metals from unbounded MSWI-BAA was affected by the leaching metal species, especially in the long term. This is consistent with Saikia et al. [42] who also pointed out the nature of the host phase is important for heavy metal leachability.

Regarding the leaching process, Zn, Cu and Pb leaching values remained relatively stable in the second half of the experiment; however, Cr kept increasing with contact time, which is really a slow process. Although contact time affects the leaching of heavy metals, the time for attaining the solid and liquid phase equilibrium is related to the metal species leaching. Specifically, the leaching data of Cr showed that equilibrium may not be attained even with 28 days of leaching, given the increasing trend. Haykiri-Acma et al. [43] also reported a similar finding—that the increase in leaching time led to enhanced Cr leaching, but the natural rain water they used had an initial pH value of around 5.5, which is lower than the one used in this study.

The leaching process of these heavy metals is probably due to the dissolution of chemical constituents in MSWI-BA and the microstructure of MSWI-BA in addition to the type of heavy metal. It is reported [42–45] that MSWI-BA has various secondary minerals, such as calcium silicate hydrate gel and ettringite, and can form various surface complexes of hydrous oxides and silicates when it contacts water. These secondary minerals and surface complexes can be easily dissolved in water and incorporate metallic ions through chemical and physical processes, including ion exchange, physical adsorption and chemical adsorption. These physical and chemical processes can be promoted by MSWI-BA particles due to their porous microstructure and irregular shape with rough surface texture [44,45]. However, the heavy metals adsorbed on an MSWI-BA surface can release again when conditions change (e.g., pH value, temperature or contact time). Therefore, the leaching of heavy metals from MSWI-BAA is a slow and complicated process, showing a fluctuation in leaching concentration throughout the experiment. Unfortunately, the influence of the porous microstructure of MSWI-BA on the leaching could not be fully explained here, due to the lack of scanning electron microscopy (SEM) analysis on the MSWI-BAA samples before and after the leaching tests.

Not surprisingly, an obvious difference was found in leaching levels between the HVEP test and simulated leaching experiments for each heavy metal, as shown in Figure 8. For all MSWI-BAA samples, most of the Zn concentrations in the simulated experiment were higher than those in the HVEP test, which is opposite to what was shown for Cr and Cu. However, most of the Pb concentrations in the simulated leaching experiment were between the highest and lowest leaching levels in the HVEP test. Although both tests were conducted under the same experimental conditions (e.g., temperature, L/S ratio, leachant and initial pH), the objectives were different. The HVEP test was actually an accelerated process to evaluate the maximum toxicity of MSWI-BAA, and the simulated leaching experiment was used for investigating the leaching under realistic conditions. It is said that different types of leaching tests give different results but reflect different aspects of leaching behavior [42]. Thus, both standard and simulated methods are important and necessary for better understanding of leaching behavior of heavy metals from MSWI-BAA.

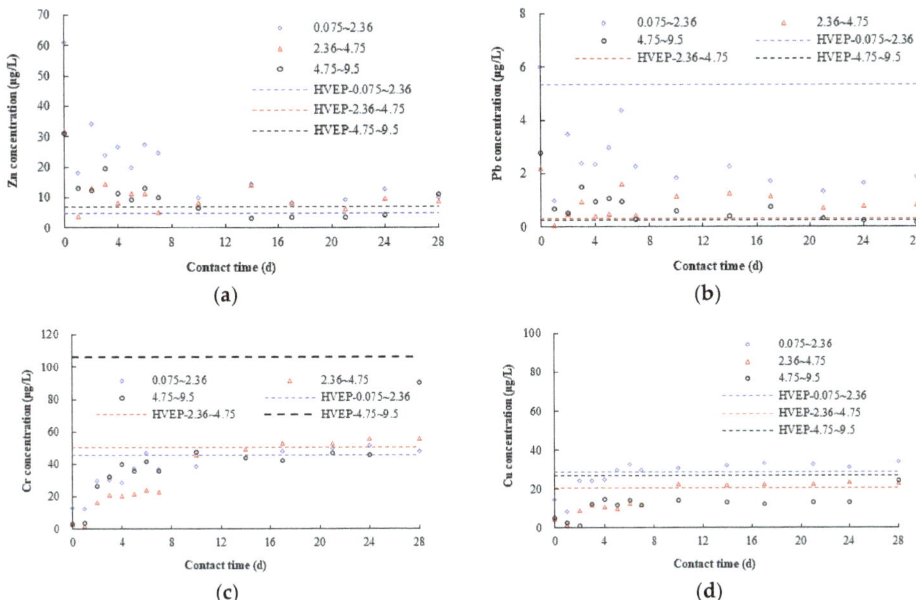

Figure 8. Leaching concentrations of heavy metals from MSWI-BAA by both the HVEP test and simulated leaching experiment: (**a**) Zn; (**b**) Pb; (**c**) Cr and (**d**) Cu

4.3. Leaching of Heavy Metals from PAC-13 containing MSWI-BAA

The interesting part of the experiment conducted on the PAC-13 specimens is that the leaching concentrations changed with contact time in some cases, for specimens containing MSWI-BAA. The outcomes were similar to those data from the control specimen (as shown in Figures 6 and 7). This might suggest that the leaching of each heavy metal might be governed by the same process, regardless of whether PAC-13 contains MSWI-BAA or not. Huang et al. [32] reported a similar result but the specimen they used was a conventional, dense asphalt mixture with different percentages of MSWI-BAA (0, 25, 50, 75 and 100%).

For three PAC-13 specimens containing MSWI-BAA, the leaching concentrations of heavy metals were affected by their species because significant differences were identified in leaching concentrations among three PAC-13 specimens with MSWI-BAAs of 0.075–2.36 mm ($p = 0.00033$), 2.36–4.75 mm ($p = 0.011$) and 4.75–9.5 mm ($p = 0.022$) in the one-factor ANOVA analysis. Significant differences were also found in leaching concentrations among Zn ($p = 0.0013$), Cu ($p = 0.00016$) and Pb ($p = 0.010$) in terms of their leaching behaviors, indicating that MSWI-BAA particle size plays a role in heavy metal leaching when used as substitute material for a natural aggregate in PA mixtures. Tasneem et al. [34] explained that when MSWI-BAA is combined with a natural aggregate in PA mixtures, the leaching of heavy metals is not only affected by the type of leaching metal and MSWI-BAA particle size, but also affected by the combined effect of the PA mixture and MSWI-BAA.

To further analyze the influence of an asphalt binder on heavy metal leaching, leaching data from both the unbounded MSWI-BAA and PAC-13 mixture containing MSWI-BAA were used to create Figure 9. The leaching values of metals from PAC-13 specimens containing MSWI-BAA showed different patterns from the unbounded MSWI-BAA. Specifically, the leaching concentrations of Zn and Pb from PAC-13 with MSWI-BAA were generally higher than those from unbounded MSWI-BAA, and the leaching process was more stable throughout the experiment. However, the leaching concentrations of Cr from PAC-13 with MSWI-BAA were generally lower than those from unbounded MSWI-BAA, and it showed a more slowly increasing process. Moreover, unlike Zn, Pb and Cr, the leaching of Cu from PAC-13 with MSWI-BAA was stable over time. A similar leaching pattern for Cu and Zn was reported by Tasneem et al. [34] using HMA, but with different contents of MSWI-BAA and for unbounded MSWI-BAA. Huang et al. [32] also reported a similar finding for Cr and Cu, but the leaching time was 10 days, which is shorter. One of the potential reasons for these results is asphalt encapsulation. The asphalt binder in a mixture forms an immobilizing film on the rough surface of MSWI-BAA particles as a surface cover, which can effectively isolate an MSWI-BAA's contact with water, thereby decreasing element mobility [21,42]. The compaction for preparing specimens is also considered an important factor affecting metal leaching [32]. In addition, the leaching might be chemically varied due to pyrolysis and the varied polarity of asphalt [34], which needs further study.

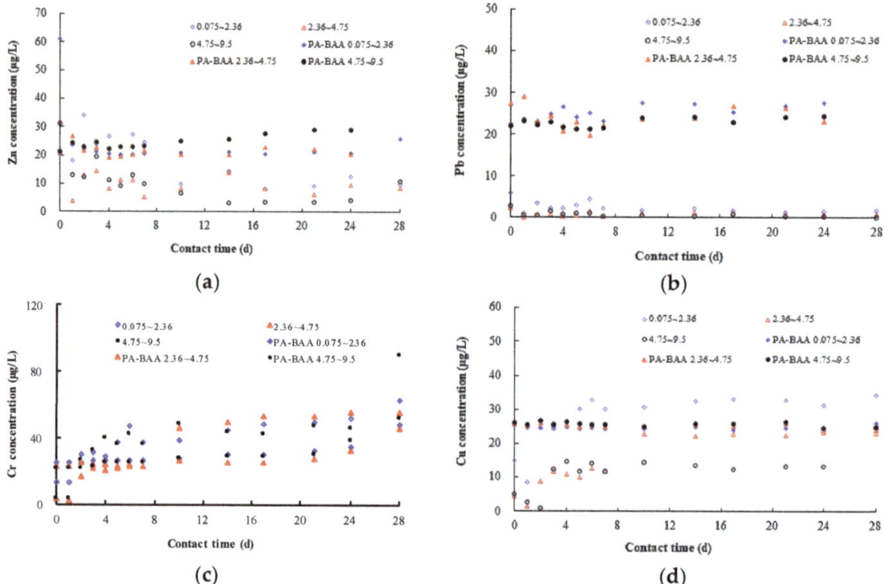

Figure 9. Leaching concentrations of heavy metals from unbounded MSWI-BAAs and PAC-13 specimens containing MSWI-BAAs: (**a**) Zn; (**b**) Pb; (**c**) Cr and (**d**) Cu.

4.4. Environmental Risk Evaluation

To assess the direct environmental impact of using MSWI-BAA in a PA mixture on surrounding water and to explore the feasibility of recycling effluent discharging from PA with MSWI-BAA for irrigation purposes, all leaching data were compared to the regulatory limit values in the Chinese standards for surface water (GB 3838-2002) [46], subsurface water (GB/T 14848-2017) [47] and irrigation water (GB 5084-2005) [48].

Table 2 shows the comparison of leaching concentrations from the HVEP test to limit values in the standards. As shown in this table, all the leaching concentrations of Cr, Cu, Zn and Pb match the limit values recommended for Class III subsurface water, Class II surface water and irrigation water, which means that these leachates can be water sources for drinking, industrial and agricultural purposes, without considering other requirements. In this sense, these four heavy metals in MSWI-BAA do not pose a short-term risk of toxicity to surrounding waters when MSWI-BAA is directly exposed to stormwater for several hours.

Similarly, the leaching concentrations of Cu, Zn and Pb in the simulated experiment were far below the corresponding limit values listed in Table 2. However, Cr concentrations exceeded the limit of 50 µg/L for Class III subsurface water and Class II surface water after 17 days of leaching, which means that these leachates are not suitable for drinking and industrial purposes, but safe for irrigation. Specially, the highest Cr concentration measured at the last sampling event was almost 1.8 times the limit value, and it would be worse over time due to its increasing tendency. These results indicate that, although the leaching of Zn, Pb and Cu from MSWI-BAA over a long period of time has little direct effect on surrounding water quality, the leaching of Cr should be of great concern.

For PAC-13 containing MSWI-BAA, all Zn and Cu concentrations were below the limit values listed in Table 2. All Pb concentrations exceeded the limit of 10 µg/L for Class III subsurface water and Class II surface water, but were below the limit for irrigation water. Cr concentrations were below the limit values for Class III, Class II and irrigation water, except for the last sampling point. These results indicate that these leachates cannot be the water source for drinking and industrial purposes, but they are safe for irrigation. However, similarly to unbounded MSWI-BAA, the leaching of Cr from PAC-13 containing MSWI-BAA should of great concern. In addition, the leaching of Pb should be given substantial attention due to the extremely high Pb concentration in leachate from PAC-13 with MSWI-BAA.

5. Conclusions

This study investigated and compared leaching behaviors for four heavy metals (Zn, Pb, Cu and Cr) from unbounded MSWI-BAA and PAC-13 mixtures containing MSWI-BAA. The HVEP test and simulated leaching experiments were conducted on MSWI-BAAs of three particle sizes (0.075–2.36 mm, 2.36–4.75 mm and 4.75–9.5 mm), but only the simulated leaching experiment was conducted on PAC-13 mixtures with and without MSWI-BAA. Leaching data were also compared to the regulatory limit values to evaluate the toxicity of leachate.

The HVEP data from unbounded MSWI-BAA showed that the type of leaching heavy metal directly affected the leaching. MSWI-BAA particle size was proven to have very good linear relationships with the leaching concentrations of Cr, Zn and Pb, which indicates that the leaching levels of Cr, Zn and Pb are strongly correlated to MSWI-BAA particle size.

Regarding the simulated leaching experiment for unbounded MSWI-BAA, the leaching concentrations of four heavy metals showed different trends over contact time. Increase in leaching time led to regular increases in Cr and Cu and regular decreases in Pb and Zn, indicating that contact time is another factor affecting heavy metal leaching.

For the PAC-13 mixture with MSWI-BAA, in general, the leaching concentration of Cu remained stable over time, Pb and Zn concentrations showed a small fluctuation and Cr showed a continuously increasing trend throughout the experiment. Similar tendencies in leaching concentrations over the contact time were also observed for the control mixture. This indicates that the presence of MSWI-BAA in a PAC-13 mixture does not necessarily change the basic tendency of heavy metal leaching over time, but it is able to lead to an increase in Cr and Zn in leachate overall.

The Cr and Cu concentrations in leachate from PAC-13 containing MSWI-BAA showed an overall decrease compared with unbounded MSWI-BAA, which suggests that the encapsulation of MSWI-BAA particles by asphalt binder is effective for reducing Cr and Cu concentrations in leachates.

All leaching data indicate that the leaching process of heavy metals was comprehensively influenced by contact time, the metal species leaching and MSWI-BAA particle size whether MSWI-BAA is used alone or mixed with a natural aggregate in PAC-13 mixtures. According to the results of environmental risk assessment, both unbounded MSWI-BAA and PAC-13 mixtures containing

MSWI-BAA have very little negative impact on surrounding surface and subsurface water quality, and the leachate is safe for irrigation. But the leaching of Cr and Pb should be monitored and mitigated.

Author Contributions: In this study, author Y.Z. designed the permeable asphalt mixture and the simulated leaching experiment; author Y.-T.Z. performed the experiments and analyzed the data; author Y.Z. wrote and edited this paper; author Y.Z. was responsible for the funding acquisition.

Funding: This research was funded by The National Science Foundation of the Jiangsu Higher Education Institutions of China, grant number 17KJB580007; The Natural Science Foundation of Jiangsu Province, China, grant number BK20170933; The Research and Development Programs of Ministry of Housing and Urban-Rural Development of the People's Republic of China, grant number 2018-K9-074, 2019-K-160.

Acknowledgments: The authors would like to thank Yu Hu and Cheng-Mao Xu for their assitance in the laboratory experiment.

Conflicts of Interest: The authors declare no conflict of interest.

References

1. Ji, Z.; Pei, Y. Geopolymers produced from drinking water treatment residue and bottom ash for the immobilization of heavy metals. *Chemosphere* **2019**, *225*, 579–587. [CrossRef] [PubMed]

2. Hjelmar, O. Disposal strategies for municipal solid waste incineration residues. *J. Hazard. Mater.* **1996**, *47*, 345–368. [CrossRef]

3. Seniunaite, J.; Vasarevicius, S. Leaching of copper, lead and zinc from municipal solid waste incineration bottom ash. *Energy Procedia* **2017**, *113*, 442–449. [CrossRef]

4. Gao, X.; Yuan, B.; Yu, Q.L.; Brouwers, H.J.H. Characterization and application of municipal solid waste incineration (MSWI) bottom ash and waste granite powder in alkali activated slag. *J. Clean. Prod.* **2017**, *164*, 410–419. [CrossRef]

5. Tang, P.; Florea, M.V.A.; Spiesz, P.; Brouwers, H.J.H. Characteristics and application potential of municipal solid waste incineration (MSWI) bottom ashes from two waste-to-energy plants. *Constr. Build. Mater.* **2015**, *83*, 77–94. [CrossRef]

6. Department for Environment Food and Rural Affairs of UK. Incineration of Municipal Solid Waste. 2013. Available online: https://assets.publishing.service.gov.uk/government/uploads/system/uploads/attachment_data/file/221036/pb13889-incineration-municipal-waste.pdf (accessed on 17 July 2019).

7. Brück, F.; Fröhlich, C.; Mansfeldt, T.; Weigand, H. A fast and simple method to monitor carbonation of MSWI bottom ash under static and dynamic conditions. *Waste Manag.* **2018**, *78*, 588–594. [CrossRef]

8. Zekkos, D.; Kabalan, M.; Syal, S.M.; Hambright, M.; Sahadewa, A. Geotechnical characterization of a municipal solid waste incineration ash from a Michigan monofill. *Waste Manag.* **2013**, *33*, 1442–1450. [CrossRef]

9. Margallo, M.; Taddei, M.B.M.; Hernández-Pellón, A.; Aldaco, R.; Irabien, A. Environmental sustainability assessment of the management of municipal solid waste incineration residues: A review of the current situation. *Clean Technol. Environ. Policy* **2015**, *17*, 1333–1353. [CrossRef]

10. National Statistical Bureau of the People's Republic of China. *Chinese Statistical Yearbook-2018*; China Statistics Press: Beijing, China, 2018.

11. National Bureau of Statistics and Ministry of Ecology and Environment of the People's Republic of China. *China's Environmental Statistic Yearbook-2018*; China Statistics Press: Beijing, China, 2019.

12. National Development and Reform Commission and Ministry of Housing and Urban-Rural Development of the People's Republic of China. National Thirteenth Five-Year Plan for the Municipal Solid Waste Harmless Treatment Facilities Development. Available online: http://www.ndrc.gov.cn/zcfb/zcfbtz/201701/W020170122611891359020.pdf (accessed on 14 February 2018).

13. An, J.; Golestani, B.; Nam, B.H.; Lee, J.L. Sustainable utilization of MSWI bottom ash as road construction materials, part I: Physical and mechanical evaluation. *Airfield Highw. Pavements* **2015**, *9*, 225–235. [CrossRef]

14. Kim, J.; Tasneem, K.; Nam, B.H. Material characterization of municipal solid waste incinerator (MSWI) ash as road construction material. *Pavement Perform. Monit. Model. Manag.* **2014**, 100–108. [CrossRef]

15. Sorlini, S.; Collivignarelli, M.C.; Abba, A. Leaching behaviour of municipal solid waste incineration bottom ash: From granular material to monolithic concrete. *Waste Manag. Res.* **2017**, *35*, 978–990. [CrossRef]

16. Del Valle-Zermeño, R.; Chimenos, J.M.; Giró-Paloma, J.; Formosa, J. Use of weathered and fresh bottom ash mix layers as a subbase in road construction: Environmental behavior enhancement by means of a retaining barrier. *Chemosphere* **2014**, *117*, 402–409. [CrossRef] [PubMed]

17. Silva, R.V.; De Brito, J.; Lynn, C.J.; Dhir, R.K. Environmental impacts of the use of bottom ashes from municipal solid waste incineration: A review. *Resour. Conserv. Recycl.* **2019**, *140*, 23–35. [CrossRef]

18. Wielgosiński, G.; Wasiak, D.; Zawadzka, A. The use of sequential extraction for assessing environmental risks of waste incineration bottom ash. *Ecol. Chem. Eng. S* **2014**, *21*, 413–423. [CrossRef]

19. Loginova, E.; Volkov, D.S.; Van De Wouw, P.M.F.; Florea, M.V.A.; Brouwers, H.J.H. Detailed characterization of particle size fraction of municipal solid waste incineration bottom ash. *Clean. Prod.* **2019**, *207*, 866–874. [CrossRef]

20. Yong, J.Y.; Lee, J.Y.; Kim, J.H. Use of raw-state bottom ash for aggregates in construction materials. *J. Mater. Cycles Waste Manag.* **2019**, *21*, 838. [CrossRef]

21. Chen, J.; Chu, P.; Chang, J.; Lu, H.; Wu, Z.; Lin, K. Engineering and environmental characterization of municipal solid waste bottom ash as an aggregate substitute utilized for asphalt concrete. *J. Mater. Civ. Eng.* **2008**, *20*, 432–439. [CrossRef]

22. Hassan, H.F. Recycling of municipal solid waste incinerator ash in hot-mix asphalt concrete. *Constr. Build. Mater.* **2005**, *19*, 91–98. [CrossRef]

23. Luo, H.; Chen, S.; Lin, D.; Cai, X. Use of incinerator bottom ash in open-graded asphalt concrete. *Constr. Build. Mater.* **2017**, *149*, 497–506. [CrossRef]

24. He, J.; Valeo, C.; Bouchart, F.J.C. Enhancing urban infrastructure investment planning practice for a changing climate. *Water Sci. Technol.* **2006**, *53*, 13–20. [CrossRef]

25. Huang, J.; Valeo, C.; He, J. The influence of design parameters on stormwater pollutant removal in permeable pavements. *Water Air Soil Pollut.* **2016**, *227*, 311. [CrossRef]

26. Valeo, C.; Gupta, R. Determining surface infiltration rate of permeable pavements with digital imaging. *Water* **2018**, *10*, 133. [CrossRef]

27. Xue, H.; Yang, T.; Liu, X.; Wu, S. Analysis of heavy metal contamination in urban road-deposited sediments in Nanjing, China. *Fresenius Environ. Bull.* **2018**, *27*, 6661–6667.

28. Huang, J.; He, J.; Valeo, C.; Chu, A. Temporal evolution modeling of hydraulic and water quality performance of permeable pavements. *J. Hydrol.* **2016**, *533*, 15–27. [CrossRef]

29. Zhao, Y.; Zhou, S.Y.; Zhao, C.; Valeo, C. The influence of geotextile type and position in a porous asphalt pavement system on Pb(II) removal from stormwater. *Water* **2018**, *10*, 1205. [CrossRef]

30. Zhao, Y.; Tong, L.; Zhu, Y. Investigation on the properties and distribution of air voids in porous asphalt within relevant to the Pb(II) removal performance. *Adv. Mater. Sci. Eng.* **2019**, *2019*, 1–13. [CrossRef]

31. Ma, X.; Li, Q.; Cui, Y.C.; Ni, A.Q. Performance of porous asphalt mixture with various additives. *Int. J. Pavement Eng.* **2018**, *19*, 355–361. [CrossRef]

32. Huang, C.M.; Chiu, C.T.; Li, K.C.; Yang, W.F. Physical and environmental properties of asphalt mixtures containing incinerator bottom ash. *J. Hazard. Mater.* **2006**, *137*, 1742–1749. [CrossRef]

33. Toraldo, E.; Saponaro, S. A road pavement full-scale test track containing stabilized bottom ash. *Environ. Technol.* **2015**, *36*, 1114–1122. [CrossRef]

34. Tasneem, K.M.; Eun, J.; Nam, B.H. Leaching behaviour of municipal solid waste incineration bottom ash mixed with hot-mix asphalt and Portland cement concrete used as road construction materials. *Road Mater. Pavement Des.* **2016**, *6*. [CrossRef]

35. National Environmental Protection Administration of the People's Republic of China. *Technical Specifications on Sampling and Sample Preparation from Industry Solid Waste HJ/T 20-1998*; China Environmental Science Press: Beijing, China, 1998.

36. National Environmental Protection Administration of the People's Republic of China. *Technical Specification on Identification for Hazardous Waste HJ/T 298-2007*; China Environmental Science Press: Beijing, China, 2007.

37. Ministry of Communications of the People's Republic of China. *Test Methods of Aggregate for Highway Engineering JTG E42-2005*; China Communications Press: Beijing, China, 2005.

38. Ministry of Housing and Urban-Rural Development of the People's Republic of China. *Technical Specification for Permeable Asphalt Pavement CJJ/T 190-2012*; China Architecture and Building Press: Beijing, China, 2012.

39. Ministry of Transport of the People's Republic of China. *Standard Test Method for Bitumen and Bituminous Mixtures for Highway Engineering JTG E20-2011*; China Communication Press: Beijing, China, 2011; China Environmental Science Press: Beijing, China, 2010.

40. Ministry of Environmental Protection of the People's Republic of China. *Solid Waste-Extraction Procedure for Leaching Toxicity-Horizontal Vibration Method NJ 557-2009*; Ministry of Environmental Protection of the People's Republic of China: Beijing, China, 2009.

41. Bendz, D.; Suer, P.; Sloot, H.; Kosson, D.; Flyhammar, P. *Modelling of Leaching and Geochemical Processes in an Aged MSWIBA Subbase Layer*; Värmeforsk Service AB: Stockholm, Sweden, 2009.

42. Saikia, N.; Kato, S.; Kojima, T. Composition and leaching behaviours of combustion residues. *Fuel* **2006**, *85*, 264–272. [CrossRef]

43. Haykiri-Acma, H.; Yama, S.; Ozbek, N.; Kucukbayrak, S. Mobilization of some trace elements from ashes of Turkish lignites in rain water. *Fuel* **2011**, *90*, 3447–3455. [CrossRef]

44. Bayuseno, A.P.; Schmahl, W.W. Understanding the chemical and mineralogical properties of the inorganic portion of MSWI bottom ash. *Waste Manag.* **2010**, *30*, 1509–1520. [CrossRef] [PubMed]

45. Zhu, Y.; Zhao, Y.; Zhao, C. Experimental study on environmental safety of cement stabilized MSWI-BA macadam base. *Bull. Chin. Ceram. Soc.* **2018**, *37*, 3296–3302.

46. National Environmental Protection Administration, General Administration of Quality Supervision, Inspection and Quarantine of the People's Republic of China. *Environmental Quality Standards for Surface Water GB 3838-2002*; China Environmental Science Press: Beijing, China, 2002.

47. General Administration of Quality Supervision, Inspection and Quarantine of the People's Republic of China, Standardization Administration of the People's Republic of China. *Standard for Groundwater Quality GB/T 14848-2017*; China Standard Press: Beijing, China, 2017.

48. General Administration of Quality Supervision, Inspection and Quarantine of the People's Republic of China, Standardization Administration of the People's Republic of China. *Standards for Irrigation Water Quality GB 5084-2005*; China Standard Press: Beijing, China, 2005.

Article

The Low-Impact Development Demand Index: A New Approach to Identifying Locations for LID

Sarah Kaykhosravi [1], Karen Abogadil [1], Usman T. Khan [1,*] and Mojgan A. Jadidi [2]

[1]   Department of Civil Engineering, Lassonde School of Engineering, York University,
      Toronto, ON M3J 1P3, Canada; saherehk@yorku.ca (S.K.); karenabs@my.yorku.ca (K.A.)
[2]   Geomatics Engineering, Department of Earth & Space Science & Engineering, Lassonde School of
      Engineering, York University, Toronto, ON M3J 1P3, Canada; mjadidi@yorku.ca
*   Correspondence: usman.khan@lassonde.yorku.ca; Tel.: +001-41-6736-2100 (ext. 55890)

Received: 28 August 2019; Accepted: 6 November 2019; Published: 8 November 2019

Abstract: The primary goal of low impact development (LID) is to capture urban stormwater runoff; however, multiple indirect benefits (environmental and socioeconomic benefits) also exist (e.g., improvements to human health and decreased air pollution). Identifying sites with the highest demand or need for LID ensures the maximization of all benefits. This is a spatial decision-making problem that has not been widely addressed in the literature and was the focus of this research. Previous research has focused on finding feasible sites for installing LID, whilst only considering insufficient criteria which represent the benefits of LID (either neglecting the hydrological and hydraulic benefits or indirect benefits). This research considered the hydrological and hydraulic, environmental, and socioeconomic benefits of LID to identify sites with the highest demand for LID. Specifically, a geospatial framework was proposed that uses publicly available data, hydrological-hydraulic principles, and a simple additive weighting (SAW) method within a hierarchical decision-making model. Three indices were developed to determine the LID demand: (1) hydrological-hydraulic index (HHI), (2) socioeconomic index (SEI), and (3) environmental index (ENI). The HHI was developed based on a heuristic model using hydrological-hydraulic principles and validated against the results of a physical model, the Hydrologic Engineering Center-Hydrologic Modeling System model (HEC-HMS). The other two indices were generated using the SAW hierarchical model and then incorporated into the HHI index to generate the LID demand index (LIDDI). The framework was applied to the City of Toronto, yielding results that are validated against historical flooding records.

Keywords: low impact development; sustainable urban drainage systems; stormwater modelling; urban development; GIS; SAW; decision-making; strategic planning; spatial analysis

---

1. Introduction

Increased urbanization has led to a significant increase of impervious surfaces. This has led to higher levels of stormwater runoff, resulting in higher flood risks, overflowing sewer systems, and damage to existing stormwater infrastructure [1]. This situation is made worse by the impacts of climate change, causing more intense rainfall events and droughts, which threaten urban water security. Social implications suggest that these risks will disproportionately affect more vulnerable populations [2].

Low impact development (LID) has become one of the most popular methods for managing stormwater and mitigating floods [3,4]. Examples of LID include bioretention cells, green roofs, detention tanks, and permeable pavement systems [5–10]. From a water and stormwater management perspective, the main purpose of most types of LID is runoff mitigation [4,11–15]; although multiple environmental and socioeconomic benefits of LID also exist [16]. Additional environmental benefits include improved water quality [10,12,17–20], decreased air pollution [21–26], and enhanced biodiversity [14,21,27–32]. Socioeconomic benefits [33,34] include provisioning services [35], educational improvements [36], enhanced immune function [37–39], and enhanced aesthetics. Not all types of LID provide all additional benefits. The types which provide these additional benefits are generally known as green infrastructure (GI) [4,40]. The term GI is a concept that goes far beyond stormwater and outside this context [4,40]. GI is a term related to landscape architecture and urban planning and focuses on creating and connecting natural ecosystems and greenway corridors (e.g., forests and floodplains), by which numerous environmental and socioeconomic benefits are provided [4,40]. In the context of stormwater, GI refers to engineered-as-natural ecosystems, such as bioretention cells, green roofs, and tree-based GI, such as tree boxes [5–10,40] (which are also LID practices) [40]. Other types, such as underground tanks (also known as cisterns), mostly provide hydrological (flood control) benefits and are typically known as LID infrastructure (not GI). Despite the fact that LIDs such as cisterns do not provide indirect benefits and have maintenance issues [41] their high efficiency in runoff management preserves them as one of the LID types [42,43].

During the strategic planning stage of stormwater or flood management projects, decision-makers select sites for LID under limited financial resources [44]. Since LID benefits are spatially dependent, maximizing these benefits requires careful planning when selecting sites for LID implementation [45,46]. This can be done through the use of geospatial data such as topography, precipitation, land cover, and multiple physical phenomena (e.g., hydrological and hydraulic processes). Without a systematic geospatial framework, selecting sites that can maximize the benefits of LID (need or demand for LID) becomes difficult. Indeed, in a comprehensive review of spatial allocation of LID, Kuller et al. [47] determined that the lack of such a geospatial framework is one of the biggest gaps in this field.

1.1. Existing Geospatial Decision Models

Factors for determining sites for LID can be grouped into two categories: feasibility—which refers to sites where LID can be implemented; and demand—which refers to sites wherein the needs or demands for the benefits of LID are high [46,47]. In the literature, three types of LID geospatial decision-making models or frameworks exist. The first prioritizes different hydraulic solution scenarios using stormwater models (SWM), which allows for modelling LIDs [48]. The second conducts geospatial analysis using geographical information systems (GIS)-based framework. The third combines the use of both SWM and GIS.

The first type of decision-making models use SWM along with a multi-criteria decision analysis (MCDA), or multi-attribute decision-making (MADM) [49]. In these models, multiple criteria are often used to identify optimum hydraulic solutions in predetermined locations; e.g., [50–54]. In these types of studies, the method for determining which sites are more effective to include are based solely on the results of the SWM and not through geospatial analysis (e.g., Lee et al. [55] used software to determine the best GI solution).

Models that conduct geospatial analysis through a GIS framework have mostly focused on determining feasible sites, and outputting maps for each type of LID. The criteria used to determine feasibility are often based on specific technical guidelines (e.g., suitable slope or distance to existing infrastructure), and rarely consider maximization of LID benefits (an example of this type of study is [56]). Kuller et al. [46] proposes a decision-making framework to determine the feasibility and demand for LID. For the demand (which is the also the focus of the present research), they selected five criteria which were grouped in three categories, including provisional (proximity to water demand), regulatory (heat vulnerability, a connected impervious-area, and floods), and cultural (visibility). However, the criteria selected for determining demand does not account for the hydrological-hydraulic demand for LID, nor is this demand integrated with the other benefits (i.e., provisioning, regulating, and cultural) of LID. Thus, there is an opportunity to expand on a previously proposed framework (such as [46] or others highlighted below) to incorporate a physically-based approach to capture the impacts of the hydrological-hydraulic demand of LID.

The third type of decision-making model framework combines both GIS and SWM. Models of this type identify sites based on feasibility and prioritize them according to either a specified decision-making system (MCDA model) or discrete prioritization (MADM model); the former may include geospatial analysis while the latter typically does not. Examples of criteria that have been used include slope, water table, hydrological soil group, and runoff volume [57]. Others also consider stakeholder opinions and technical experts [54]. For example, one study [58] did consider three classes of benefits of LID (the environmental, economic, and social benefits), though these were done independently (i.e., in isolation of each other) rather than using a holistic approach. With these three different classes of decision-making criteria, identical rankings of the sites were obtained [58]. Thus, as concluded by Kuller et al. [47], although a set of GIS-MCDA tools and frameworks have been developed, none of them are sufficient and comprehensive. In addition, in another review by Lerer et al. [59], they concluded that there is a lack of a comprehensive model that covers all physically-based aspects (referred to as "How Much models") and human aspects (referred to as "Where and Which" models).

1.2. Geospatial, Physically-Based Framework

Amongst previous studies in this field, the strategical allocation of LID and prioritizing sites on this basis, particularly through considering hydrological and hydraulic parameters, is less frequently addressed. Of those that have addressed it, multiple limitations remain that have significant environmental and socioeconomic impacts. One recent study by Martin-Miklea et al. [60] used the concept of hydrologically sensitive areas (HSA) [60]. Since HSA was originally used for pollution transport risk [61], the HSA procedure and the criteria used do not match the physical processes that occurs in LID. For example, in their study [60], the slope is considered to have an inverse impact on the priority of a site in terms of the need for LID, meaning that sites with higher slopes have lower priority for installing LID (their study included rain-barrel, green roof, porous pavement, rain garden, vegetated swale, detention and retention pond, and riparian buffer). This is incorrect from a hydrological perspective. Sites with high slope generate a lower time of concentration (t_c). Many LID types (such as grass swales) can be used to increase the t_c (and hence reduce runoff related issues). Though types of LID without outflow do not affect the t_c (e.g., drainage boxes), capturing the runoff of a high slope site prevents the runoff from combining with runoff (from an adjacent site) with low t_c to the downstream runoff. Thus, steeply-sloped sites should have high priority (or high demand) for implementing LID for runoff management. Also, the method proposed by Martin-Miklea et al. [60] does not consider the geospatial or temporal distribution of rainfall. In hydrological models, the importance of the spatial and temporal distribution of rainfall has been noted by many researchers [62]. There are several studies (such as [63]) that present different methods for accurate modelling of rainfall distribution over watersheds. Thus, for including the spatial rainfall distribution over the study area, particularly large-scale methods are necessary. In addition, the geospatial distribution of land cover and impervious area was also neglected by Martin-Miklea et al. [60]—in urban areas a significant portion of the natural land-cover is modified. This modification of the land-cover is important to be considered if the study area is under the influence of human activities and has experienced land-cover changes. Moreover, the approach presented in Martin-Miklea et al. [60] has multiple mathematical limitations. For example, in their study, to avoid mathematical errors, for sites where the hydraulic conductivity was zero, the index was reassigned somewhat arbitrarily to -3. This value of -3 is a number that is lower than the highest index calculated in the case study. Therefore, this number is a replacement for the actual calculations and depends on the case study and needs justification for each specific case. Finally, in the Martin-Miklea et al. [60] study, only the hydrological-hydraulic demand or need for LID is considered, whilst the environmental and socioeconomic benefits are largely ignored.

Overall, while previous studies have developed GIS-MCDA frameworks to determine the feasibility or demand of LID, there is a need to develop an integrated framework for the strategic planning stage of LID projects to determine where to install LID [47] based on the "demand" or "need," specifically, by integrating hydrological and hydraulic processes with high demand, due to other indirect benefits. In previous studies, prioritizing sites based on their potential to generate stormwater runoff was one of the least addressed subjects. In addition, prior studies either overlooked integrated hydrological-hydraulic benefits with other indirect benefits of LID (i.e., socioeconomic or environmental status [64]), or used inadequate types and numbers of assessment criteria [46] (e.g., the hydrological-hydraulic benefits or indirect benefits were overlooked). Also, many of previous studies were based on an SWM model. In these types of studies, the scale of the study area is limited to the SWM itself, which limits the extent of the study area. This restriction is due to the complexity (in terms of required input data) of developing SWMs for large-scale regions. Thus, as recommended by other studies, such as [12,59,65,66], more investigations to address the indirect benefits of LID should be integrated into GIS-MCDA models for LID to identify the demand for LID, which could include future climate, land-use, and socioeconomic scenarios.

Therefore, the objective of the present research was to develop a physically-based geospatial decision-making framework to identify the demand for LID in urban areas. The proposed framework is conceptually similar to and builds on existing frameworks (e.g., [46] and [60])—as geospatial data is used to determine LID demand. However, the proposed framework introduces three new indices and two new heuristic relationships to determine LID demand which integrates both the direct and indirect benefits of LID. The first heuristic relationship (which uses the proposed hydrological-hydraulic index, HHI) is based on hydrological and hydraulic principles, focuses solely on runoff quantity and not quality (building on the framework by Martin-Miklea et al. [60]), and uses actual physical values of hydrological and hydraulic data in the overlaying process (i.e., the data are not rescaled). Since this represents the physical processes that occur within LID, the HHI is validated against results from a physically-based model. This relationship ranks the sites based on their potential for runoff generation (represented by the HHI). Next, a GIS-MCDA framework was developed using two additional indices (the environmental index, ENI and socioeconomic index, SEI) to rank the sites based on their demand for LID based on indirect benefits. From the stormwater management perspective, this research considers runoff control as the primary benefit of LID. A second innovative heuristic relationship was then developed to integrate the indirect benefits of LID (represented by the ENI and SEI) with the HHI (the main benefits of LID). This relationship integrates the indirect benefits as an additive value to the direct benefits of LID, meaning that if HHI is not zero (there is a hydrological-hydraulic demand), the indirect benefits (ENI and SEI) add value to the HHI to estimate the demand for LID holistically. But, if HHI is zero (i.e., no runoff-related demand for LID) and the indirect benefits are not zero, the overall demand for LID is estimated to be zero.

In our framework, there are no mathematical limitations, and actual values of all parameters can be used as inputs. Finally, unlike models relying on the SWM, the proposed framework has no limits on the scale of the study area, since the number of input data are independent from the scale of the area, and thus, the only cost is the computational time, which can increase based on the size of the area.

This paper is organized as follows: Section 2 presents the materials and methods, including the case study; geospatial data; the development of three new indices—the hydraulic-hydrologic index (HHI), the environmental index (ENI), and the socioeconomic index (SEI); and an approach to combine those indices into the proposed LID demand index (LIDDI). Section 3 includes the results and discussion, including the generation of the HHI, ENI, and SEI maps for the case study and the validation of the HHI, along with the final LIDDI map. Section 4 follows with the conclusions.

2. Materials and Methods

To address the inadequacies of existing approaches for the spatial allocation of LID, a geospatial, physically-based framework is proposed, which is then combined with a multi-criteria decision-making model. This framework was developed to identify sites with the greatest demand or need for LID, and to maximize both the direct and indirect benefits of LID.

To do this, three indices were developed to consider the direct (stormwater runoff reduction) and indirect (environmental and socioeconomic) benefits of LID. The first index, HHI, was developed based on the physical principles governing stormwater runoff. HHI ranks the sites based on their potential for runoff generation. The second and third indices are ENI and SEI, respectively, which were generated using a SAW method within a hierarchical decision-making model. ENI and SEI identify sites with the highest demands for LID in terms of environmental and socioeconomic needs. Combining these three indices, the LID demand index (LIDDI) was then produced. The LIDDI ranks all sites within the study area based on their demand or need for LID based on hydraulic-hydrological, environmental, and socioeconomic benefits.

The conceptual framework of the method described above is presented in Figure 1. In this figure, the variables represent the physically-based parameters (for HHI) and geospatial criteria or indicators (for ENI and SEI). These variables, the secondary and primary indices, are described in detail in the following sections (Section 2.3 to Section 2.5).

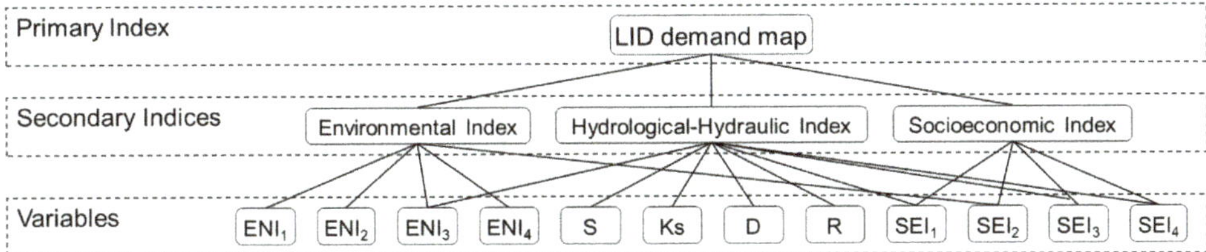

Figure 1. The schematic of the conceptual framework of low impact development (LID) demand index (LIDDI) generation, where S is slope (degree), K_s is saturated hydraulic conductivity (mm/hr), D is depth to restrictive layer (mm), and R is rainfall intensity (mm/hr).

2.1. Study Area

The utility of the proposed framework is demonstrated through a case study: the study area used for this research was the City of Toronto, which is located in Southern Ontario, Canada, and covers an area of 630.2 km^2 with a population of approximately 2.73 million. The land-use varies from residential, commercial, and industrial to green spaces. Eleven major rivers pass through the city, dividing the city into eleven watersheds. Figure 2 presents the watersheds and their intersections with the administrative border of the City of Toronto. The use of LID has been growing in Toronto over the last several years [67]. Toronto is experiencing a transition period from grey (end of pipe methods) to green (using LID) infrastructure. This is being done through developing and implementing a range of policy instruments, by-laws, and regulations, which include developing and re-designing existing policies associated with land-use, the environment, water, infrastructure, and planning [15]. Some examples include the Green Roof By-law and the Downspout Disconnection By-law [15].

2.2. Geospatial Data

In this study, data from open, publicly available sources were used and are listed in Table 1. Those geospatial data were used to derive each of the parameters used for three indices; the rationale for selecting each dataset is discussed in detail below, for each index. Each dataset was converted into raster format with a ground resolution of 5 × 5 m (to align them with the digital elevation model, DEM, resolution which was used as the reference resolution for this research). The framework was developed and implemented in ArcGIS.

The proposed framework, and specifically, the application presented in this research, is restricted by the resolution of available data. Here, a raster resolution of 5 × 5 m was used due to the available DEM data. Higher or lower resolutions can be chosen; however, this should be done with intent, as it can impact the accuracy and reliability of the results.

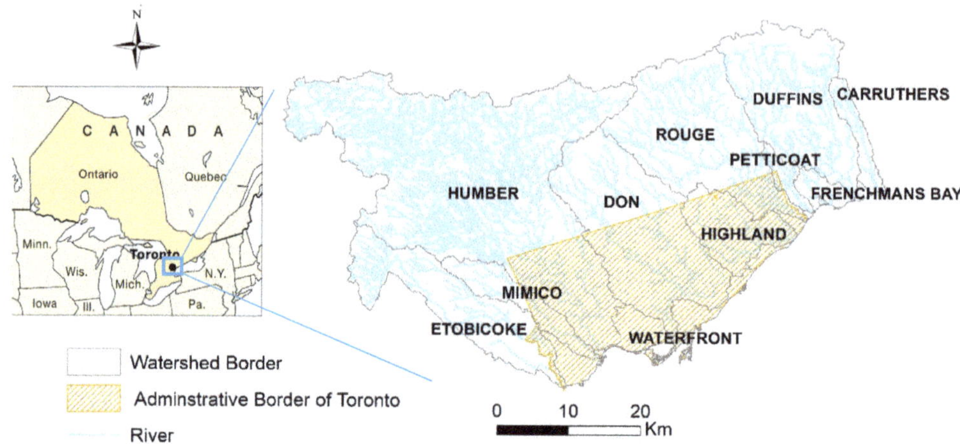

Figure 2. The extent of the study area (hatched in orange color), which is the administrative boundary of the City of Toronto and its intersection with the eleven watersheds in Toronto.

Table 1. The raw data downloaded from publicly-available sources, including the structure/geometry of the data, and the source of the data.

ID	Data	Data Structure/Geometry	Data Source
1	Air pollution ($PM_{2.5}$)	Vector/point	Ministry of Environment, Conservation and Parks
2	Bedrock layer	Vector/polygon	Ministry of Northern Development and Mines
3	Census tract boundaries 2016	Vector/polygon	Statistics Canada
4	Demographics data-Census 2016	Tabular data	Statistics Canada
5	Digital elevation model (DEM)	Raster/5 m × 5 m	Ontario Ministry of Natural Resources and Forestry-Provincial Mapping Unit
6	Ecozones of Ontario	Vector/polygon	Scholars Geoportal
7	Groundwater table	Vector/point	Groundwater Information Network (GIN)
8	Land cover	Raster/0.6 m × 0.6 m	City of Toronto
9	Toronto rivers	Vector/polyline	Toronto and Region Conservation Authority (TRCA)
10	Surficial geology and saturated hydraulic conductivity (K_s)	Vector/polygon	Government of Canada open data website
11	Toronto precipitation data	Vector/point	Government of Canada open data website
12	Educational institute	Vector/point	City of Toronto
13	Hospital	Vector/point	Google Earth

Note: Missing data were ignored in the calculations.

2.3. Hydrological-Hydraulic Index (HHI) Development

HHI ranks the cells (i.e., the pixels of the raster data) by their runoff or flood-generation potential. Cells with higher HHI generate higher volume and peak runoff flow rate (referred to as high-runoff cells). These cells should be the first target for implementing LID. However, where it is not feasible to implement LID, the next target should be selected based on several parameters, including the topography of the catchment, amongst other parameters [51]. The selection of the feasible sites is out of scope of this research and requires another investigation which incorporates results of this study (demand) to the feasibility.

To generate HHI, the geospatial variables that are representative of the runoff hydrograph were identified and quantified. Then, a geospatial overlaying heuristic relationship (Equation (1)) was developed according to the physical principles. The variables were then spatially overlain, based on the relationship proposed (Equation (1)).

To identify the variables, the runoff generation process and the associated relationships (mathematical equations) were considered. Four variables were identified and quantified for HHI: rainfall intensity (R), hydraulic conductivity (Ks), water storage capacity of soil (D), and catchment slope (S).

Runoff is dependent on rainfall and infiltration. Rainfall intensity, denoted by R and measured in mm/hr, is especially important in large scale study areas where uneven spatial distribution of rainfall can be an important factor. By a simplification assumption we neglected the temporal distribution of R and it was considered a constant. The value of R was obtained from the intensity-duration-frequency (IDF) curve for the City of Toronto and represents an extreme rainfall event. For this research, the intensity of the 100-yr, 5-minute duration event was selected as R to represent extreme events. This rainfall is a sample of a low probability and intense rainfall event, though any other R value may be used, since the focus is on the difference of R across the study area (rather than the magnitude itself).

Infiltration is a function of the soil moisture, porosity, hydraulic conductivity, and water storage capacity of the soil. The Green-Ampt equation [68] demonstrates that if the antecedent soil moisture is saturated, the infiltration rate is equal to the saturated hydraulic conductivity (K_s). This allows one to estimate the infiltration rate using K_s data: cells with higher K_s values have higher infiltration rates, and consequently, generate less runoff. Note that urban areas contain significant impervious surfaces, whose K_s value is essentially zero, resulting in an infiltration rate that is also zero. Thus, the land-use data (specifically the impervious surface layer) was combined with the K_s of surficial geology data using the minimum geospatial operation tool in ArcGIS. With the minimum operation, each cell the K_s of impervious surfaces (assumed to be 0 mm/hr) and the K_s of the surficial geology layer (which is greater than or equal to 0 mm/hr) was compared and the lower value was assigned to each cell. In addition, streams, rivers, and other waterbodies within the study area were excluded from K_s layer and potential LID sites. Through geospatial subtraction operation, K_s values were subtracted from the rainfall intensity, R, resulting in an indicator for the runoff generation potential for each cell.

Infiltration volume is also dependent from the water storage capacity of the soil, which is further dependent on soil characteristics (e.g., soil porosity and moisture) and the thickness of the soil layers. Capacity, denoted by D, is quantified through three variables: the depth to groundwater table (D_g), the depth to the first impermeable soil layer (bedrock) (D_r), and the soil porosity (n). The value of D takes the minimum of either D_g or D_r, and represents the distance between the soil surface and the first restrictive layer. Capacity is then multiplied by the soil porosity, n, which quantifies the porous storage volume in the vertical direction. The calculated volume allows for the additional quantification of runoff generation potential that considers both the infiltration rate and the water storage capacity. For instance, a cell with a high infiltration rate can be restricted by an impermeable soil layer, resulting in decreased infiltration.

The time of concentration (t_c) is another contributing variable to runoff generation; LID can be used to increase t_c in urban areas to reduce damage resulting from stormwater runoff. Two cells with the same values of K_s, R, and D can generate the same runoff volume; however, the t_c may be different. Using Kirpich's formula [69] t_c can be estimated using the catchment length and slope (S). When using cells with equal areas, the t_c is a non-linear exponential function of the slope to the power of 0.385. To represent this nonlinear relationship of t_c and S, we generated the slope layer from the DEM data in ArcGIS and estimated the $S^{0.385}$ value using for each 5 × 5 m cell.

By identifying the contributing variables (R, K_s, D, and S), we developed a heuristic equation (Equation (1)) to geospatially overlay the runoff-contributing variables. Our relationship was developed based on the theoretical concepts of runoff generation. The validation of this relationship is presented in Section 3.4. Using this equation, a value for each cell is calculated based on the potential of cells to generate runoff. The cells within the study area can then be compared in terms of their need or demand for LID. Cells with higher HHI have higher demand for LID; installing LID in a cell with a higher HHI captures more runoff or prevents the generation of runoff at that site, and attenuates t_c:

$$\text{HHI}_j = \begin{cases} \left[R_j - \left((K_s)_j \cap (K_i)_j\right) \cap \left(n_j \times \left((D_g)_j \cap (D_r)_j\right)\right)\right] \times \left[1 + (\tan S)^{0.385}\right] & \text{for } R_j > Ks_j \\ 0 & \text{for } R_j \leq Ks_j \end{cases}, \tag{1}$$

where j is the cell number; HHI_j is the HHI at cell j; R_j is rainfall intensity (mm/hr) at cell j; $(K_s)_j$ is the saturated hydraulic conductivity of the surficial geology layer (mm/hr) at cell j; $(K_i)_j$ is the hydraulic conductivity of impervious areas (e.g., roads, parking lots, and building roof tops) at cell j, set equal to zero for impervious cells; n_j is the soil porosity at cell j; $(D_g)_j$ and $(D_r)_j$ are, respectively, depth to groundwater (mm) and depth to the first restrictive soil layer (mm) at cell j; and S is the terrain slope (degrees) at cell j. Note that since physics-based variables were used to formulate this heuristic equation, dimensional consistency was ensured for all parameters considered. Also, with a lack of soil data, it can still be used. Whereas the data for K_s and D do not exist, there are two ways of using the equation. Either K_s and D can be assum based on known characteristics of the study area or assuming the worst-case scenario, which is value of zero for these parameters ($K_s = 0$ and $D = 0$).

To generate the HHI map for the City of Toronto, the four variables in Equation (1) (R, K_s, D, and S) were generated from the available raw data (Table 1) and converted to raster layers. Rainfall intensity, R, was derived from the IDF tables of three meteorological stations within the City of Toronto: Oshawa, Oakville Southeast, and Toronto Buttonville. The raster rainfall data for the study area were generated by geospatial interpolation using the point data from the three meteorological stations. The hydraulic conductivity, K_s, was derived from the surficial geology and land cover data as explained above (Table 1). To generate rainfall storage capacity of the soil (D), D_g was generated using groundwater level data of wells (in point data format). Using the geospatial interpolation within the study area and groundwater level data from the wells, the raster data layer of D_g was produced. D_r was generated from the bedrock data (in polygon data format). Using these two variables and a minimum geospatial operation tool in ArcGIS, the raster map of D was generated. The soil porosity (n), was assumed to be 0.5 for the entire study area due to the lack of available porosity data. According to Chow et al. [70], this is an average estimate of porosity for different soil types (sand, loam, silt, and clay). Terrain slope (S) was derived from DEM data using the slope tool in ArcGIS. In addition to these variables, a data layer consisting of the existing LID (using only existing tree canopy was considered which was derived from the land-cover data) was also generated in order to eliminate these sites as potential sites for implementing LID. Due to the lack of data of actual LIDs currently installed within the study area, other types of LID, such as green roofs, raingardens, etc., were neglected in this study. Due to this, the existing LID facilities are still considered potential LID sites; however, if this data were available, existing LID sites can be eliminated from the analysis if required.

2.4. Environmental Index (ENI) Development

The ENI quantifies the relative demand of cells for LID (especially GI) in terms of environmental needs. ENI is associated with mitigating or preventing environmental damage that results from anthropogenic sources, such as urbanization. The environmental criteria were selected based on the DPSIR (drivers, pressures, state, impact, and response) model developed by the European Environment Agency [71]. Based on DPSIR, urbanization is the "driving force" causing "pressures" that include pollution, and changes in land-use and habitats' populations. These "pressures" cause changes to the "state" of the environment (air quality, biodiversity, water quality, and soil quality). These changes lead to negative "impacts" on public health and ecosystems, which can elicit a positive or negative response feeding back into the "driving forces" (or to the "impacts" and "states" directly). GI is considered a "response" model of intervention to these "impacts"; other "responses," such as environmental laws can also be applied. The criteria for ENI represent the environmental "states"; the four sample environmental states used for this research were: air quality, bio-habitat, water quality, and soil quality. Specifically, the criteria included air pollution denoted by ENI_1, biodiversity denoted by ENI_2, water quality denoted by ENI_3, and soil contamination denoted by ENI_4, all of which are caused by urbanization and for which GI can be used as response for impact mitigation.

Higher ENI_1 values were assigned to sites with higher concentrations of air pollution, as they were considered to receive the greatest benefit from the GI in terms of air quality [21–24]. In particular, LID (planted-based types) reduces the concentrations of a variety of air pollutants that are known to increase health risks, such as fine atmospheric particulate matter ($PM_{2.5}$) [72,73]. In the proposed framework, $PM_{2.5}$ concentration was chosen as the indicator of overall air pollution, as is often done in the literature.

Higher ENI_2 values were assigned to sites with higher biodiversity to ensure the existing bio-habitats were maintained. GI provides potential bio-habitats, and consequently, prevents the impact of urbanization on existing habitats [14,21,27–32]. For example, previous studies have shown that green roofs can host diverse plants and animals, as well as function as stepping stones to other nearby habitats, particularly for immobile organisms [74]. In another study, this effect was highlighted when a 41% increase in forest coverage led to an increase of 81 bird species—highlighting that tree-based LID or GI may have a positive impact on bird species [75]. Thus, the ecozones of Ontario (collected from the Scholars Geoportal website, http://geo2.scholarsportal.info) were used as indicators of the biodiversity within the study region.

GI is intended to capture and treat stormwater at the source (often referred to as "source control"). In doing so, GI can effectively capture contaminants (e.g., sediment which itself contains other chemicals, such as metals) [36]. Previous research has shown that GI such as bioretention cells can treat stormwater through physical, chemical, and biological processes. Specifically, it provides filtration, sedimentation, adsorption, and plant and microbial uptake [76]. Through these processes, pollutant concentrations of metals (zinc and copper), total suspended solids, total phosphorus, and total nitrogen have been reduced by more than 90% [76–78]. Capturing these contaminants at the source (i.e., upstream of the drainage point or outlets to rivers) prevents the contaminants from flowing downstream and also reduces the spread of contaminants within the watershed [79–81]. Thus, for this research, higher ENI_3 values were assigned to sites farthest from rivers as an indicator of "source control" to improve water quality in the rivers. Areas closer to the rivers were given lower priority (lower ENI_3), whereas areas farther upstream were given higher priority in terms of LID (especially GI) need or demand.

Sites with GI installed maintain neutral pH, increased metal retention, and reduced risk of diffusion, all of which are attributed to the enhancement of biochemical processes in the soil by GI [19,82–84]. To prevent the contamination of soil, stormwater runoff and sites closest to pavements were assigned higher ENI_4 values (and thus higher priority). Roads and parking lots were considered to be the sources of the soil contamination yielding high concentrations of pollutants such as zinc, copper, lead, and cadmium, most often found at inflow points [77].

For the City of Toronto, the ENI_1 values were derived from point data of eleven air-quality monitoring stations from the Ministry of Environment, Conservation, and Parks. The air quality data layer was interpolated and converted to a raster layer using an inverse-distance-weight method. The ENI_2 values were derived from the total population of terrestrial bird, mammal, reptile and amphibian, and tree species per unit area of the ecozone (from the ecozones of Ontario database). The ENI_3 were calculated using the Euclidean distance, river cells have distance equal to zero and cells farther from the river have higher Euclidean distances. For ENI_4 values, road and parking lot surfaces were extracted from the land cover map, assigned as sources, and distances from these sources were estimated using the Euclidean distance similarly to the ENI_3 process.

All four selected criteria were standardized by applying the linear standardization method (Equation (2)).

$$SENC_{j,k} = \frac{ENC_{j,k}}{(ENC_k)_{max}}, \tag{2}$$

where k is the total number of criteria (in this case, 4), j is the total number of cells in the study area, $SENC_{j,k}$ is the standardized environmental criteria k at cell j, $ENC_{j,k}$ is the actual value of environmental criteria k at cell j, and $(ENC_k)_{max}$ is the greatest possible value of environmental criteria k within the study area. After standardization, the overall ENI for each cell was calculated by overlaying all four criteria, using the simple additive weighting (SAW) method (shown in Equation (3)).

$$ENI_j = \sum_1^k w_k \times FENC_{j,k} \tag{3}$$

where ENI_j is the value of ENI at cell j; $FENC_{j,K}$ is the finalized environmental criteria k at cell j, which can be derived from Equation (4); and w_k is the weight of criteria k.

$$FENC_{j,k} = \begin{cases} SENC_{j,k}, & \text{for direct distance criteria} \\ 1 - SENC_{j,k}, & \text{for inverse distance criteria} \end{cases} \tag{4}$$

2.5. Socioeconomic Index (SEI) Development

The SEI quantifies the LID demand of each cell in terms of the socioeconomic benefits of LID (especially GI and planted-based LID). In addition to the hydrological and environmental benefits, GI also improves socioeconomic status in a variety of ways, including increasing access to green spaces, enhancing academic performance in schools, public health, and social resilience [33–35,85–90]. To address these benefits, priority was assigned to cells using four sample socioeconomic factors: population density (SEI_1), distance to green spaces (SEI_2), distance to educational centers (SEI_3), and distance to hospitals (SEI_4).

Higher SEI_1 values were assigned to sites with a higher population density in order to serve a higher population with the indirect benefits of GI. The second criteria (SEI_2) prioritizes sites that are farthest from existing green spaces; this creates an even distribution of green spaces. Areas with sparse landscaping, more deprived neighborhoods, and low public transport availability are identified as having the greatest demand for the benefits associated with LID (specially GI types) [38,88–90].

The third criteria (SEI_3) gives priority to sites closest to educational centers and schools. Studies have shown that LID and GI improve the educational performance, and socioemotional, behavioral, and cognitive development of students [91,92]. In particular, preschoolers at educational risk were found to have greater development of independence and social skills when enrolled to schools with higher levels of plant-based LID [91]. The fourth criteria (SEI_4) assigns priority to sites that are closer to hospitals to improve the health status of patients. Studies have shown that patients with green window views recovered significantly faster and required less medication compared to patients with urban views [93]. Indeed, green spaces are widely associated with enhanced immune function and greater mental well-being [39,90]. Furthermore, green spaces have been shown to lower fasting blood-glucose levels in both adults and children, effectively lowering the risk of diabetes, which is now recognized as one of the leading causes of illness and death [89].

A fifth socioeconomic benefit is social resilience, which comes in the form of provisioning services. Provisioning services yield products such as firewood, medicinal plants, and wild foods. Studies have shown that poor households and vulnerable populations highly rely on GI to support their daily livelihood needs. One study found that 57% of households on average use GI as a coping strategy as a response to co-variate shocks such as floods [87]. This benefit is harder to quantify due to the lack of available data. Because of this, quantifications of socioeconomic benefits often underestimate the actual benefits that can be obtained. Nevertheless, social resilience is highlighted as a great benefit to particularly vulnerable populations and could be included in future studies if relevant data is available.

For the City of Toronto, SEI_1 was estimated based on the 2016 census data, by dividing the population with the area of the corresponding census tract. SEI_2, the distance from existing green areas, was estimated by the Euclidean distance; higher values were assigned to sites further away from existing green areas. The raster map of the third criteria (SEI_3) was also estimated by calculating the Euclidean distance, with the locations of educational centers (in point data) assigned as sources. SEI_4 was also derived with the same process as SEI_2 and SEI_3.

As with ENI, the variables were standardized by applying the linear standardization method using Equation (5).

$$SSE_{j,m} = \frac{SEC_{j,m}}{(SEC_m)_{max}},$$ (5)

where k is the total number of criteria (in our study case, 4), j is the total number of cells in the study area, $SSEC_{j,m}$ is the standardized socioeconomic criteria k at cell j, $SEC_{j,m}$ is the actual value of socioeconomic criteria k at cell j, and $(SEC_m)_{max}$ is the greatest possible value of environmental criteria k within the study area. After standardization, the overall SEI for each cell was calculated by overlaying all four criteria, using the simple additive weighting (SAW) method (shown in Equation (6)).

$$SEI_j = \sum_1^m w_m \times FSEC_{j,m},$$ (6)

where SEI_j: is the value of SEI at cell j, $FSEC_{j,m}$ is the finalized environmental criteria k at cell j, which is derived from Equation (7), and w_m is the weight of criteria k.

$$FSEC_{j,m} = \begin{cases} SSEC_{j,m}, & \text{for direct distance criteria} \\ 1 - SSEC_{j,m}, & \text{for inverse distance criteria} \end{cases}$$ (7)

2.6. LID Demand Index (LIDDI)

The LIDDI represents the sites ranked by their respective demand for LID, with the combined consideration of all hydrological and hydraulic, environmental, and socioeconomic variables. High LIDDI values represent sites with the highest demands or needs for LID. To generate the LIDDI, the SEI and ENI were first geospatially overlaid using the SAW method to generate the combined socioeconomic-environmental index (SEENI) (Equation (8)). The relationship between the weights of the method are shown in the Equation (9). In this equation, W_{EN} is the sum of w_k, and W_{SE} is the sum of w_m.

$$SEENI_j = w_{EN}ENI_j + w_{SE}SEI_j$$ (8)

$$W_{EN} + W_{SE} = \sum_1^k w_k + \sum_1^m w_m = 1,$$ (9)

where $SEENI_j$ is the environmental-socioeconomic index at cell j, ENI_j is the environmental index at cell j, SEI_j is the socioeconomic index at cell j, w_{EN} is the corresponding weight of ENI, and w_{SE} is the corresponding weight of SEI. As demonstrated in Equations (2)–(8), the proposed framework to generate SEENI is not restricted by the number of environmental and socioeconomic criteria selected nor the assigned weight to each criterion. The number of criteria is considered, and the weights assigned to each criterion can be customized based on an end-user's needs, data availability, or other requirements. Thus, this framework has been developed in a way to be customizable to determine the need or demand for LID based on which benefits need to be considered or incorporated into the framework.

The resulting SEENI is proposed as an additive value, where it can be overlain onto the HHI through a mathematical multiplication operation to generate the LIDDI, the overall demand for LID (Equation (10)).

$$LIDDI_j = HHI_j \times (1 + SEENI_j),$$ (10)

where $LIDDI_j$ is the demand of cell j for LID.

SEENI was considered as an additive value due to the multiple advantages it provides. First, two cells with identical HHIs but differing SEENI values can be ranked distinctly based on their environmental and socioeconomic demands for LID. This allows the main objective of this framework to be achieved: spatial allocation of LID with the combined considerations of hydrological and hydraulic, environmental, and socioeconomic factors. Table 2 shows a simple example to illustrate these concepts: the effect of SEENI on the HHI and final LID demand rank of the three sample cells. Two areas with equal HHIs but unequal SEENI values (cells 1 and 2), will result in different LIDDIs, and the final rank is impacted by the cells with higher SEENI values. Secondly, since the primary goal of LID is stormwater runoff management, a site that does not generate runoff does not require LID (even if an environmental or socioeconomic demand for it exists, as shown in Table 2, cell 3). In the framework, cells with HHI values equal to zero but high SEENI values, will result in LIDDI values of zero. This ensures that LID is not unnecessarily implemented in sites without a hydrological or hydraulic need.

Table 2. The effect of the socioeconomic-environmental index (SEENI) on the hydrological-hydraulic index (HHI) in two sample cells.

Cell #	HHI	SEENI	LIDDI	Rank
1	0.9	1	1.8	1st
2	0.9	0	0.9	2nd
3	0	1	0	3rd

2.7. Adaptability of the Proposed Framework

The proposed framework can be adapted in various ways for the intended application, project requirements, or stakeholder opinions.

Our proposed framework can be used for all types of LID, generally categorized as detention (such as tanks), planted infiltration (e.g., raingarden), and dry infiltration (e.g., infiltration trenches) types [94]. Note that for our case study, we only used planted infiltration or GI types of LID for Toronto to demonstrate the utility of the approach proposed. Accordingly, all the associated co-benefits are those benefits that correspond to GI or planted infiltration LID techniques. If a different subset of LID is used while using the proposed approach, the co-benefits should be selected to be compatible with the actual co-benefits of those LIDs.

To generate HHI, we did not standardize the data, and actual values with consistent dimensions were used. Regarding the SEI and ENI, we standardized all criteria using the linear value scaling method ranging between 0 and 1 (Equations (2) and (5)). This selection was based on several reasons: the exact correlation between each single criterion and result (the decision which is going to be made) is unknown. Despite that we know that each criterion has an increasing or decreasing influence on the results, the exact linearity or nonlinearity of this change is not clearly proven. However, our framework allows the end-user to select among other standardization methods, such as score range, value function, utility function, probabilistic, and fuzzy. This selection, though should be based on the expert judgment (as it was in [46]), which was recommended and performed in studies such as [95,96].

HHI in our proposed framework allows for estimating the relative rank of sites within multiple study areas and comparing them. However, the ENI and SEI (so the LIDDI) only allow for ranking sites within one study area. The reason is that criteria for developing HHI was not standardized and actual values of data were used, whereas for ENI and SEI, all criteria were standardized. To cope with this and to compare different study areas, we investigated the possibility of using absolute global minimum and maximum values for each ENI and SEI criterion with respect to standardized versions. Standardization of data with respect to a global absolute value can allow the comparison of LIDDI values of different study areas. For the minimum value we found zero to be a rational value on a global scale, and we applied it in the framework. However, our results demonstrated that finding a global maximum value for most of criteria requires further research which is beyond the scope of the present work and should be focused on in future work.

The weights used in the SAW hierarchical model for ENI and SEI in this research were chosen based on the particular study area, team judgement, and expertise. However, as these weights are area adjustable, we suggest generating the weights through a systematic process, such as the analytical hierarchy process (AHP) using weighting methods such as the pairwise comparison (PC) combined with a method like Delphi, which relies on a panel of experts.

Also, the four criteria used for ENI and SEI, respectively, are not restricted to the criteria we selected for this study: they can be customized according to the environmental and socioeconomic concerns of the study area, and availability of data. They can be estimated and incorporated into HHI using Equations (2) to (8). Note that future work should consider the correlation between the primary and secondary benefits, as well as between different secondary benefits. For example, biodiversity and existing green spaces, or impervious areas or land-use and population density may be highly correlated and their impacts should be quantified.

3. Results and Discussion

To evaluate the effectiveness of the proposed framework, the three indices HHI, ENI, and SEI were generated, followed by the SEENI and then the LIDDI for the City of Toronto. The resulting maps are color-coded to show the gradient from lowest (green) to highest (red) demand of LID. Additionally, the HHI map was compared to the historical flood data and further validated against a physically-based hydrological model, the Hydrologic Engineering Center-Hydrologic Modeling System model (HEC-HMS) developed by the U.S. Army Corps of Engineers [97]—a commonly used hydrological model. The results are presented, analyzed and discussed in this section.

3.1. HHI Map

The resulting HHI map is presented in Figure 3f, along with the data layers of the required variables R, K_s, D, and S. The map of existing LID is included as well. Figure 3a shows that, overall, the City of Toronto has a low K_s (and therefore a high LID demand) across the city, though there are some high K_s areas in the eastern and western parts of the city. Figure 3b indicates a relatively low slope for the city, with higher slopes located along river banks. Sites with slopes greater than 45 degrees angles were excluded from potential LID sites. This was due to two reasons: first, the maximum cut slope that is stable long-term is 45 degrees [98]; second, to cope with the DEM data error, in which the difference between the elevations of some new urban developments (such as bridges) and surrounding areas were falsely estimated as a high slope. Figure 3c shows that D is higher at the same regions as high Ks, except for northern part of the city. Figure 3d shows a 20 mm/hr difference in average rainfall intensity across the city. Figure 3e shows the areas with existing tree canopies, which were eliminated from potential areas for LID implementation and excluded from further analysis. As described earlier, tree canopies were used as a proxy for existing GI due to the insufficient data available.

By geospatial overlaying of all variables using Equation (1), the HHI map was generated, as presented in Figure 3f. Comparing the HHI map Figure 3f with the input variable maps (Figure 3a–e), high HHI areas are located in areas with low hydraulic conductivity (K_s), high precipitation intensity (R), low depths to restrictive layers (D), and high terrain slopes (S).

Assigning HHI values to an area allows filtering or ranking sites by their runoff generation potential. The identification of high HHI sites targets the main source of runoff generation for LID implementation, and thereby effectively maximizes runoff reduction. Furthermore, the HHI map can show the relationship between historically flood-prone areas and high HHI sites.

To demonstrate the application of filtering cells with high HHI, we considered the highest 2.3% of HHI (mean value plus two standard deviations of HHI) cells as the highest priority for LID implementation within the region. For Toronto, these are sites with HHI values of 0.56 or greater (if we linearly rescaled HHI in range 0–1). Figure 4 presents the map of these highest 2.3% HHI cells and also contains TRCA's data on historically flood-vulnerable areas [99], shown in dashed circles. The impact of upstream flood sources (i.e., areas with high HHI) on downstream areas (the encircled flood-vulnerable areas) is evident, illustrating the usefulness and utility of the proposed HHI. This is especially critical since significant parts of the upstream Humber and Don Rivers are outside the study area, and the contribution of these areas on downstream flood generation must also be considered.

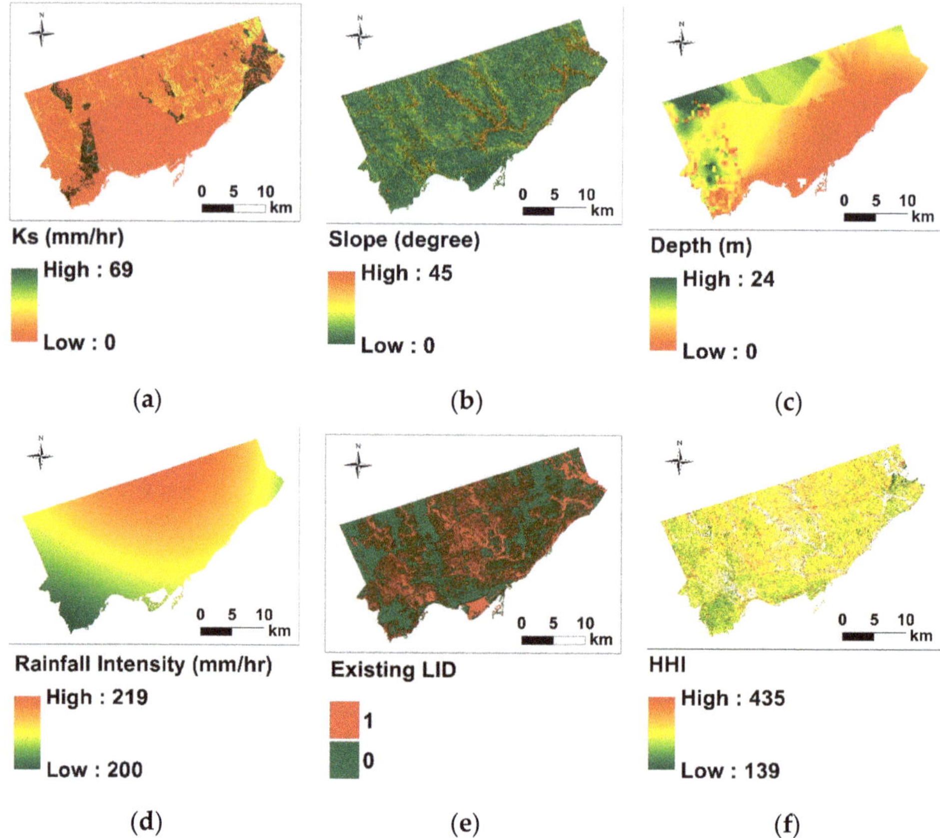

Figure 3. Input variables: (**a**) hydraulic conductivity, K$_s$; (**b**) terrain slope, S; (**c**) depth to restrictive layer, D; (**d**) rainfall intensity, R; (**e**) existing LID; and (**f**) the hydrological-hydraulic index (HHI) map generated (in all maps, red color indicates high LID demand and green color indicates low LID demand).

From Figure 4, areas upstream of the flood-prone one are often, though not always, associated with higher HHIs. Downstream of the confluence of the Don River branches, Figure 4 areas (a) and (b) (East Don, West Don, and Taylor/Massey Creek) are historically vulnerable to flooding. HHI values suggest this vulnerability originates from the upstream East Don River and the main inlet of Taylor/Massey Creek. However, high HHI areas of the western branch of the Don are lower and are mostly concentrated on the central parts of the city, and may be the origin of the flooding at Figure 4 areas (a) and (c). In addition, upstream of areas (d) and (e) shows a high concentration of high HHI values.

West of the Don watershed, the Humber watershed shows flood-vulnerable areas along and slightly east of the Black Creek River. High HHI sites are also seen along this branch (upstream of areas (f), (g), and (h)), decreasing downstream towards the confluence with Albion Creek (between areas (h) and (i)). This is in contrast to the lower HHI areas along Albion Creek (downstream of the area (k)), though there is a concentration of high HHI cells at the upstream and near (k).

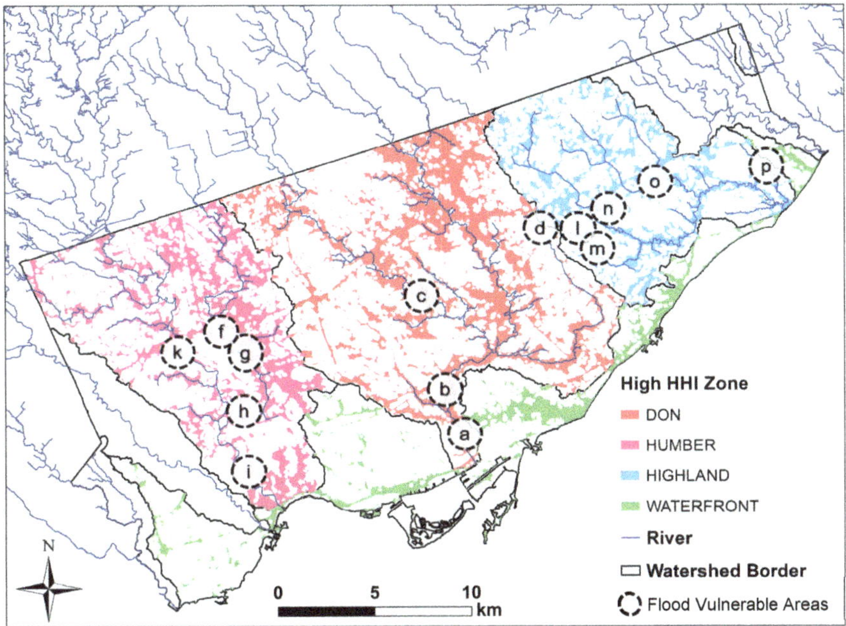

Figure 4. Top 2.3 percent ranked sites with a high standardized hydrological-hydraulic index (HHI) of 0.56 and greater (these areas are presented for each watershed within the study area) and the location of historically flood-vulnerable areas of the City of Toronto.

On the eastern part of the city, the Highland watershed also has flood-vulnerable areas located along all four branches of the main river (Malvern, Bendale, Dorset Park, and Centennial Creek), as marked by (l), (m), (n), (o), and (p) in Figure 4. HHI results shows that upstream of each of these four branches are areas with high values of HHI, and therefore, sources of runoff generation. The resulting HHI map combined with the historically flood-vulnerable areas of the city demonstrates that the HHI can successfully identify areas with high runoff generation and its relationship to downstream flood-vulnerable clusters. Thus, the HHI is effective in correlating the geographical locations of potential sources of flood to observed flood locations.

The HHI results were then statistically analyzed for the entire study area and for each main watershed. Segmenting Toronto by cells (5 by 5 m in size) resulted in 25.4 milion cells. By extracting the cells without available data, water bodies and LID 17.9 million existing, potential LID cells remain. The mean calculated, standardized HHI (if linearly rescaled between 0 and 1, which is not necessary and we only performed it for comparison purposes) for all cells is 0.4 with a standard deviation of 0.079. Figure 5 shows a plot of the HHI values versus the normalized count of cells within the entire study area and each watershed. By investigating the watersheds individually, Figure 5 demonstrates that the Don has the highest mean HHI of 0.427, whereas Etobicoke has the lowest mean HHI of 0.374. Thus, in terms or runoff generation potential, the HHI values for each watershed can show the relative demand for LID in each watershed. In this case, the demand for LID in the Don (a highly urbanized watershed and centrally located within the city) is highest, and it is lowest in Etobicoke. Further validation and verficiation of the HHI is provided in Section 3.4, where the HHI for a site and is directly compared to runoff generation volumes from a hydrological model.

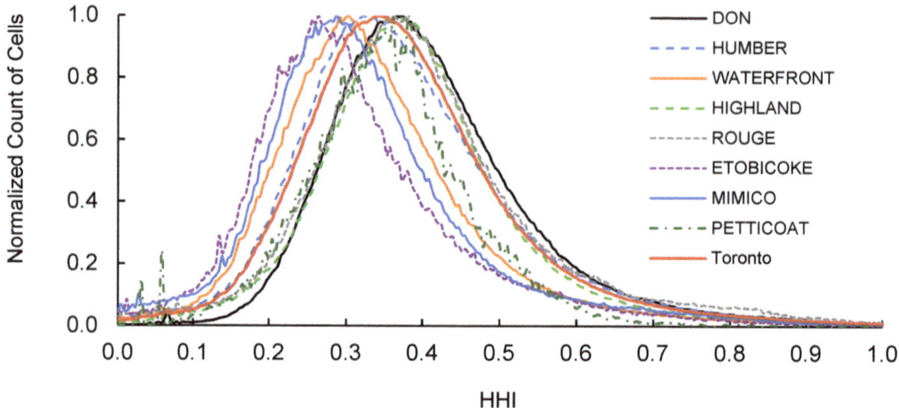

Figure 5. Normalized frequencies of the hydrological-hydraulic index (HHI) within eight watersheds and City of Toronto. Normalization of each watershed was performed by dividing the HHI frequency of each watershed by its maximum HHI frequency.

The resulting HHI from geospatial analysis of physically-based processes is a quantification of the sources of runoff generation. This allows decisions to be made about micro-scale sites (25 m^2) within a bigger macro-scale study area (630.2 km^2). HHI highlights areas that contribute the most to flooding; here, high HHI areas are shown to produce flood-vulnerable areas downstream. Selecting these sites for LID implementation improves the effectiveness of LID while efficiently allocating resources in stormwater management.

3.2. ENI Map

To generate the ENI map for the City of Toronto, data for the four criteria (ENI_1 to ENI_4) were collected, converted to raster format (gridded cells), and standardized (using Equations (2) and (3). The resultant maps are presented in Figure 6. Figure 6a shows that the concentration of air pollution (ENI_1) is located at the central areas of Toronto. ENI_2 (Figure 6b) indicates that Toronto contains four different ecozones with one being significantly more dominant, covering the majority of the city. ENI_3 (Figure 6c) shows the distance to rivers with river boundaries clearly noticeable. ENI_4 (Figure 6d) shows highly-developed areas while less developed areas are around rivers and green spaces. All four layers (ENI_1 to ENI_4) were combined using the method described in Section 2.3 in order to generate the ENI map (Figure 6e). The resultant ENI map assigned higher priorities to areas farthest from the rivers, located in the central part of the city. The highest ENI are seen in the Waterfront, Humber, Don, and western regions of the Highlands. The distribution of ENI demonstrates that distance to rivers and pavements has a higher impact on the final ENI value compared to air pollution and biodiversity; this is primarily due to the insignificant variations among of these two criteria across the study area.

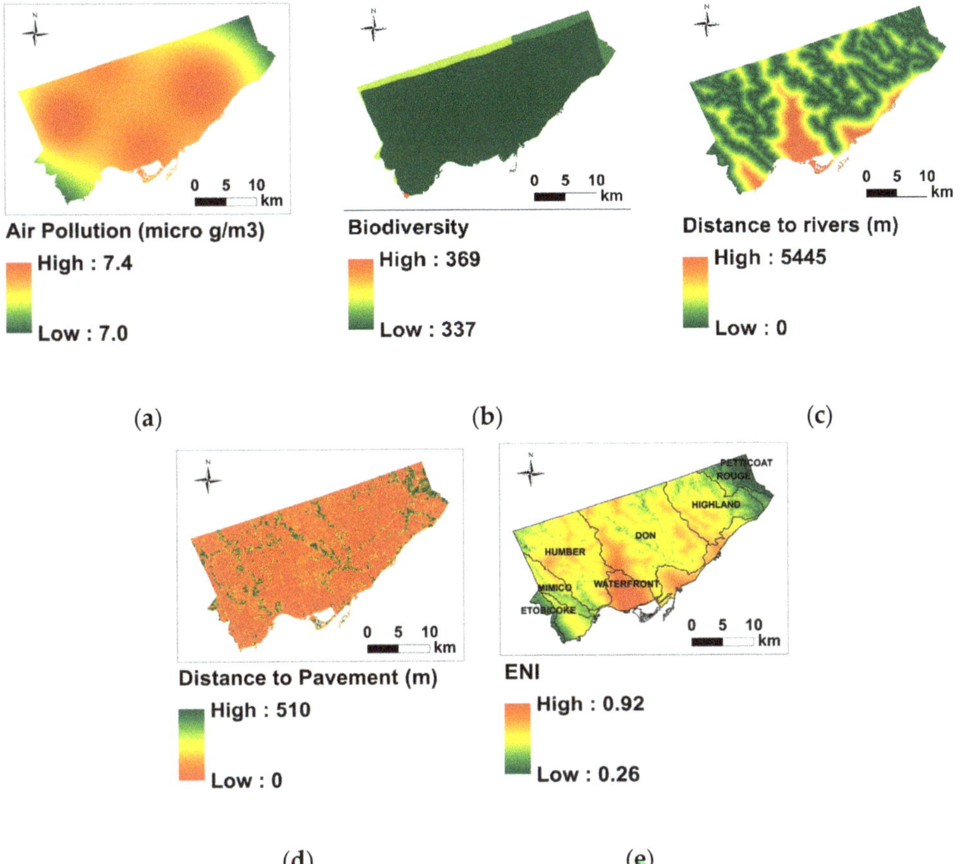

Figure 6. Input environmental variables: (**a**) air pollution (ENI$_1$), (**b**) biodiversity (ENI$_2$), (**c**) distance to rivers (ENI$_3$), (**d**) distance to pavements (ENI$_4$), and (**e**) the ENI map of City of Toronto we generated. In all maps, a red color indicates high LID demand and green color indicates low LID demand.

3.3. SEI Map

To generate SEI, all four criteria maps (SEI$_1$ to SEI$_4$) were created from the raw data, standardized, and overlain (Equations (4) and (5)). Each map is shown in Figure 7. SEI$_1$ (Figure 7a) indicates that the highest populated areas are located in Downtown Toronto. Similarly, multiple census tracts with high population densities are scattered across the city. SEI$_2$ (Figure 7b) shows that the LID demand for green spaces is higher downtown. Other high priority areas that are scattered across the city show impervious areas, some of which are used for industrial land. SEI$_3$ (Figure 7c) shows a high and evenly distributed pattern of educational institutes throughout the city. SEI$_3$ (Figure 7d) shows hospitals mostly concentrated on the south-western part of the city. All four maps that were generated were overlain to generate the SEI map (Figure 7e) with highly urbanized areas showing the greatest socioeconomic demands for LID.

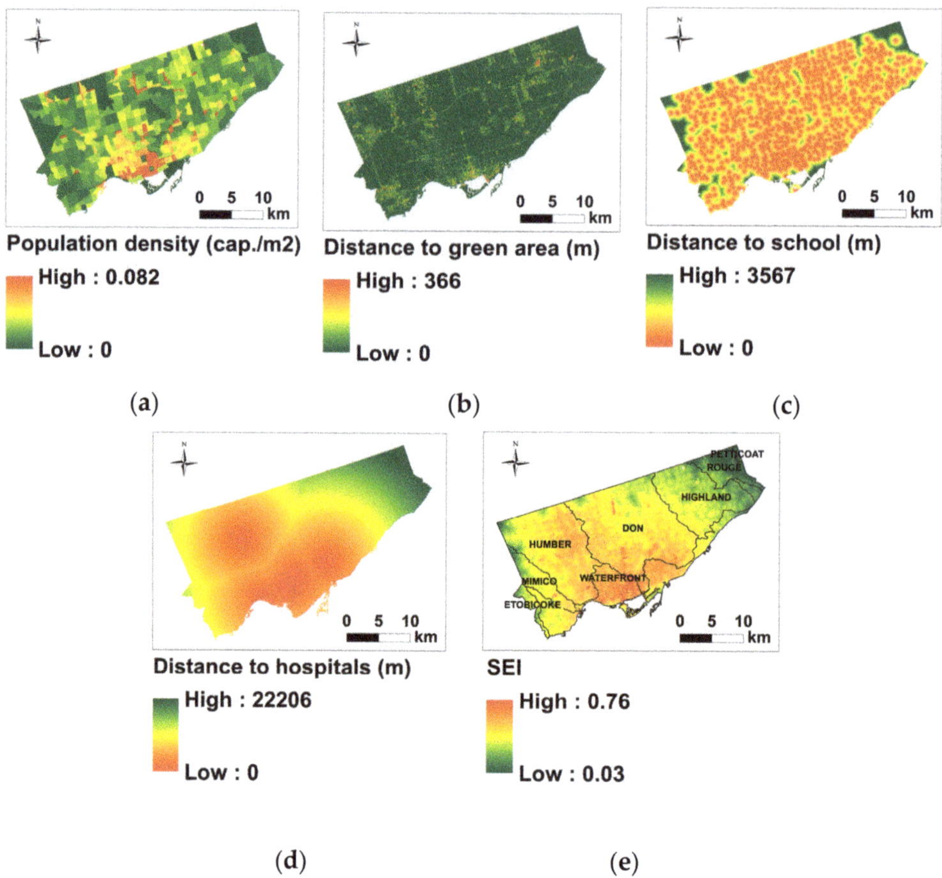

Figure 7. Socioeconomic variables: (**a**) population density, (**b**) distance to green area, (**c**) distance to educational institutes, (**d**) distance to hospitals, and (**e**) the socioeconomic index (SEI) generated. In all maps, red indicates high LID demand and green indicates low LID demand.

3.4. Validation of Indices

The HHI heuristic relationship was developed based on hydrological and hydraulic principles. HHI is a GIS-based index representing the runoff generation potential of each cell. Thus, HHI should be able to rank various catchments with different hydrological-hydraulic characteristics (R, D, Ks, and S). To examine the HHI performance, we validated the HHI against a physically-based model for the study region. Validation of ENI and SEI was not performed; the difficulties of validating these types of indices and decision-making models makes doing so not always practical since these models do not represent physical phenomena—which is discussed extensively in Kuller et al. [46].

The validation of HHI was done by comparing the results of HHI to HEC-HMS simulations. A sensitivity analysis was conducted by calculating the HHI using extreme cases for each of the variables used to develop the HHI (which are listed in Table 3) and running the HEC-HMS model for the same values. A combinatorics approach was used to define 16 variants of catchments (or scenarios) and these are illustrated in Figure 8.

Table 3. Variable values considered for 16 scenarios.

Variable	Low	High
R (mm/h)	50 [1]	220
D (mm)	0	24000 [2]
K_s (mm/h)	0.3	117.8
S (m/m)	0.0001	0.15 [3]

[1] Minimum R was not considered zero, allowing for generating a low amount of runoff. [2] The maximum D layer was selected according to maximum D of the study area. [3] The maximum S was considered to be 0.15 (=tan 9°), which is the greatest maximum proposed slope for all types of LID [57].

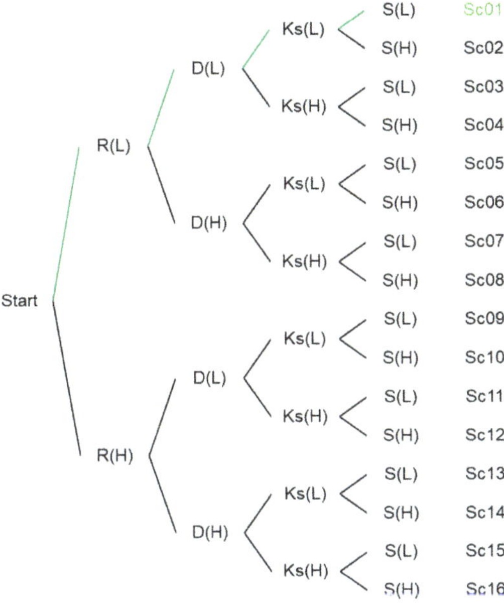

Figure 8. The 16 scenarios considered based on combinatorics concept.

The hydrographs from HEC-HMS for all 16 scenarios were created and normalized and each scenario was ranked based on the peak flow and flow volume. On the other hand, the HHIs were calculated for all 16 scenarios as well, and subsequently, these scenarios were ranked based on the HHI values. Results from both methods were compared and contrasted and are presented in Figure 9.

The methods show near identical results, with the exception of scenarios 9, 11, and 13. As expected, Sc10, Sc12, and Sc14 (high R and S, and low K_s, or D) show the highest potential for runoff generation, whereas Sc07 and Sc08 (low R and S, and high K_s and D) do not generate any runoff. HHI values are lower than the HEC-HMS values only for the flatter catchments, which was expected. Since HHI accounts for slope, runoff volume, and peak flow, the HHIs of steep catchments were expected to be higher than the HHI of catchments with low slop. Thus, the flat catchments should be considered lower priority than steeper catchments with other parameters being identical. This demonstrates how HHI is incorporates the slope with the peak flow and runoff volume in order to rank each scenario.

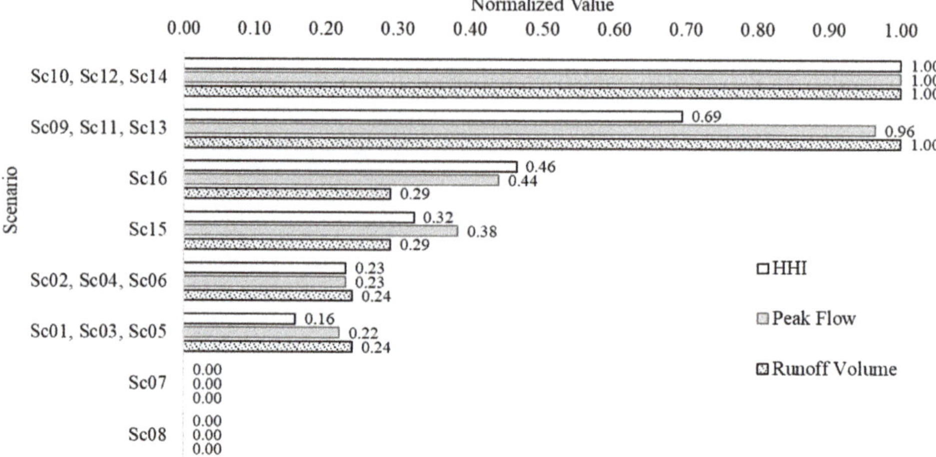

Figure 9. Comparison of HHI with the Hydrologic Engineering Center-Hydrologic Modeling System model (HEC-HMS) for 16 extreme scenarios.

The hydrographs of each catchment were compared in the following criteria: peak flow, runoff volume, and time of concentration, as shown in Figure 10. From this figure, the normalized hydrographs show approximately three types of runoff volume: high, medium, and low peaks. The rankings obtained from Figure 9 are identical to the rankings obtained from HHI. The highest peaks are also the scenarios that obtained high HHI rankings. Within each grouping, the scenarios with the steeper slopes but similar other variables are ranked higher than those with lower slopes. From the agreement between the HHI and HEC-HMS, HHI demonstrates its capability to rank scenarios based on the hydrological and hydraulic demands, using publicly-available geospatial data and geospatial analysis.

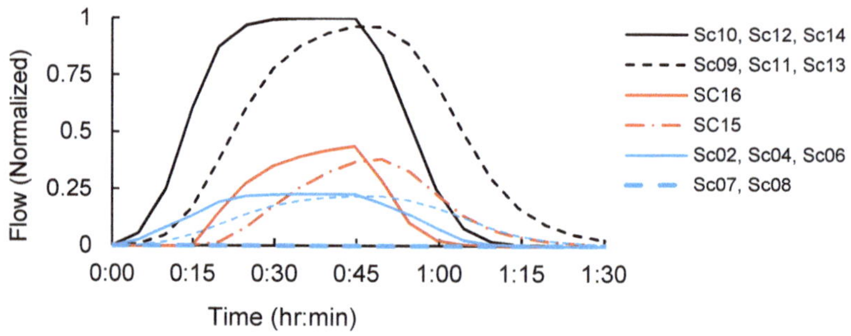

Figure 10. The normalized hydrograph of all 16 catchments resulted from modelling in the Hydrologic Engineering Center-Hydrologic Modeling System model (HEC-HMS).

3.5. LIDDI Map

To generate the LIDDI map, the SEENI was first calculated by combining SEI and ENI (Equation (6)), as presented in Figure 11. SEENI was then overlain with HHI (Equation (7)) to generate the LIDDI map, presented in Figure 12a. Values of LIDDI range from 0 to 1.62, with a mean of 0.63 and standard deviation of 0.125. The top 2.3% represent sites with values greater than two standard deviations are those with an LIDDI value of 0.88 or greater, as shown in Figure 12b.

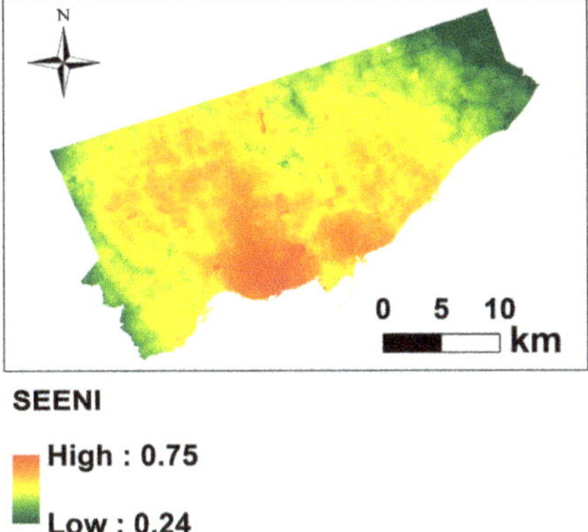

Figure 11. The generated socioeconomic-environmental index (SEENI) map resulting from geospatial overlaying of the environmental index (ENI) and socioeconomic index (SEI).

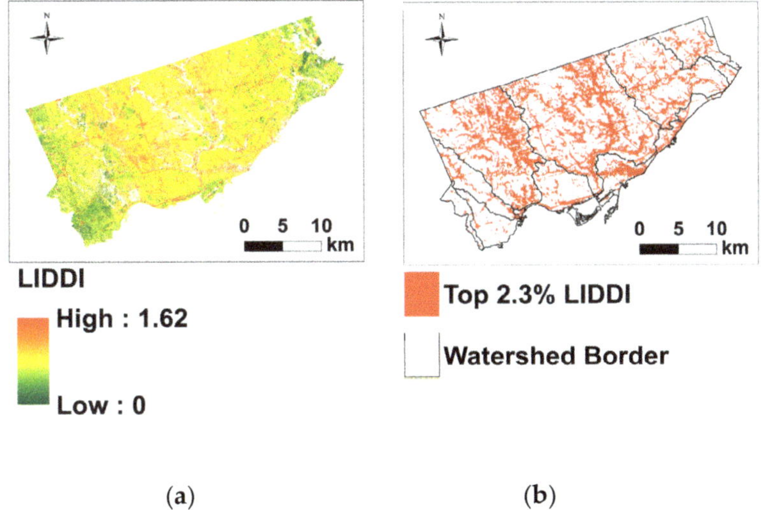

Figure 12. (a) The LID demand index (LIDDI) map generated from geospatial overlaying the socioeconomic-environmental index (SEENI) and the hydrological-hydraulic index (HHI); (b) the top 2.3 percent of sites ranked by the highest demand for LID.

By comparing the map shown in Figure 12 (b) with the HHI map shown in Figure 4, the influence of SEENI on HHI can be observed. In particular, they show the sites that were excluded (or included) due to the lack of consideration for the socioeconomic and environmental demands (i.e., the impact of SEI and ENI or overall LID demand) that would have otherwise been obtained from SEENI.

To clearly demonstrate this, two regions with low and high values of SEENI are shown in Figure 13. Figure 13 shows HHI overlain with LIDDI. In Figure 13 (c) (area with high SEENI), the cells that were previously excluded as high priority under HHI, are now included as priority under LIDDI, as shown in red. In contrast, in the low SEENI regions (Figure 13 (b)), cells that were ranked as high priority under HHI are now excluded from high LIDDI, as shown in blue.

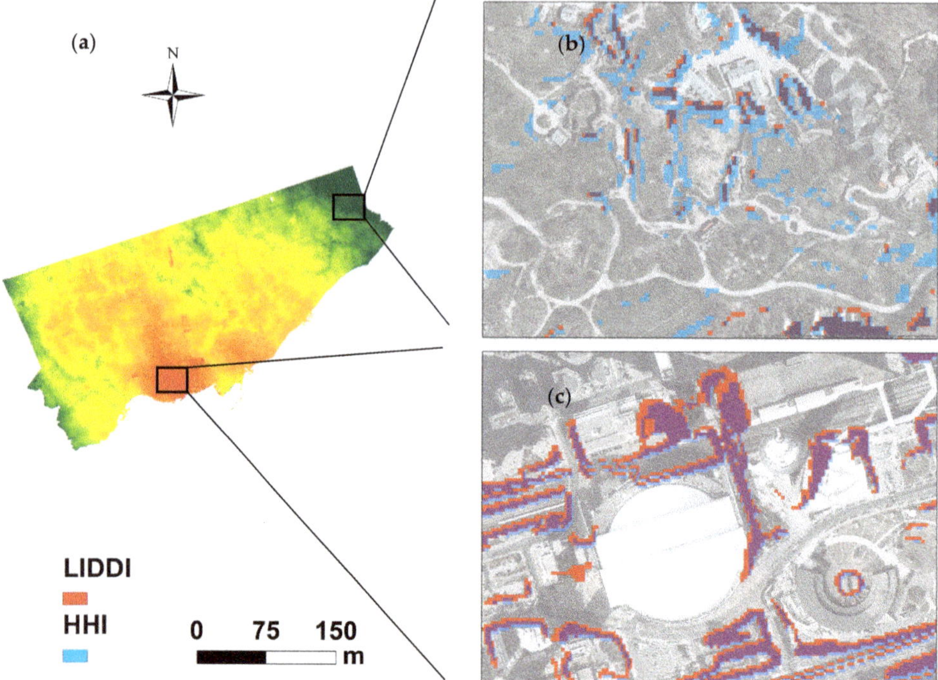

Figure 13. The effect of the socioeconomic-environmental index (SEENI) on (**a**) the hydrological-hydraulic index (HHI); and the changing in LIDDI for low and high SEENI areas: (**b**) the Toronto Zoo area (low SEENI); (**c**) the area around CN Tower in downtown Toronto (high SEENI area).

The LIDDI map allows visualization of the LID demand across a study area, which would otherwise be a complex geospatial problem. The results of this framework allow decision-makers to observe the main sources of flooding with consideration of the socioeconomic and environmental factors using a physics-based, robust, and easily-implementable index, the HHI. The distribution of LIDDI across the study area helps to visualize the need for LID in different regions in a large-scale study area. This visualization provides users with a simple expression of a complex and multi-criterion geospatial problem: it highlights the connections between the sources of runoff generation, flood prone areas, and areas that need higher socioeconomic and environmental benefits of LID. In addition, the LIDDI allows us to filter top-ranked sites for implementing LID. Thus, with limited time and financial resources, the presence of such a tool can saves our resources while it maximizes the benefits of LID.

4. Conclusions

In this research, a physically-based geospatial framework was developed to identify sites with the highest LID demand in terms of their flood generation potential, along with socioeconomic and environmental factors. To develop this framework, hydrological-hydraulic, socioeconomic, and environmental criteria were identified and generated using publicly-available geospatial data. Through geospatial analysis and the SAW method within a hierarchical decision-making model, the variables were overlaid, resulting in three different indices: HHI, ENI, and SEI. These indices rank the sites based on the demand or need for LID with respect to each index. By combining these indices, a LID demand index (LIDDI) was generated, which ranks the sites based on HHI, ENI, and SEI, and identifies locations where LID should be installed to meet the multiple objectives (hydrological-hydraulic, environmental, and socioeconomic). A heuristic equation was created to develop the hydrological-hydraulic index (HHI). To validate this equation and the HHI, we compared the results against a hydrological model (HEC-HMS) and the historically flood-vulnerable locations within the study area.

The proposed framework was applied to the City of Toronto. The results show that the framework for generating LIDDI has multiple advantages. It addresses the lack of a systematic geospatial allocation of LID methods that are based on hydraulic and hydrological principles, and includes environmental and socioeconomic factors. This helps flood mitigation strategies in future land-use planning by allowing the decision-makers to rank the sites based on their demand for LID rather than using an ad hoc approach. Furthermore, this framework was specifically developed for LID runoff management objectives, unlike some other existing frameworks, which were adapted from other objectives (e.g., water quality). Thus, many processes and benefits directly associated with LID are considered in the framework we propose. The three indices can also be used separately if only one aspect (hydrological, environmental, or socioeconomic) is needed. Moreover, publicly-available data was used for the geospatial analysis, which allows for the wide-use of our proposed framework. There is no limitation on the scale of the study area; modelling of a micro-scale, meso-scale, and macro-scale study areas are all possible. Unlike previous studies where multiple mathematical limitations exist for the physically-based index (HHI), the proposed framework does not require modification or customization to use raw input data for this index. Thus, the actual values of all input data layers can be used as input data. Finally, the proposed framework is generalized to be applicable in any study area or region of interest. The framework can be used to develop strategies for future flood risk attenuation, stormwater management, urban development and planning, retrofitting stormwater infrastructure, or developing cost-effective development plans. In particular, it supports spatial decision-making and flood-risk reduction strategies, and enhances the overall effectiveness of LID practices.

Author Contributions: Conceptualization, S.K., U.T.K., and M.A.J.; data curation, S.K.; funding acquisition, U.T.K. and M.A.J.; investigation, S.K. and K.A.; methodology, S.K.; resources, U.T.K. and M.A.J.; Software, S.K.; supervision, U.T.K. and M.A.J.; validation, S.K.; visualization, S.K.; writing—original draft, S.K. and K.A.; writing—review and editing, K.A. and U.T.K.

Funding: This research was funded by the Natural Sciences and Engineering Research Council of Canada, and York University.

Acknowledgments: The authors would like to thank ESRI Inc. for granting the student version of the ARCGIS. We would also like to thank York University Libraries for assisting with data collection. Lastly, we would like to thank the comments from three reviewers, whose feedback greatly helped improve this manuscript.

References

1. Frantzeskaki, N.; Kabisch, N.; McPhearson, T. Advancing urban environmental governance: Understanding theories, practices and processes shaping urban sustainability and resilience. *Environ. Sci. Policy* **2016**, *62*, 1–6. [CrossRef]
2. De Macedo, M.B.; do Lago, C.A.F.; Mendiondo, E.M. Stormwater volume reduction and water quality improvement by bioretention: Potentials and challenges for water security in a subtropical catchment. *Sci. Total Environ.* **2019**, *647*, 923–931. [CrossRef] [PubMed]
3. Coffman, L.; Clar, M.; Weinstein, N. Overview of low impact development for stormwater management. In Proceedings of the 25th Annual Conference on Water Resources Planning and Management, Chicago, IL, USA, 7–10 June 1998; pp. 16–21.
4. Fletcher, T.D.; Shuster, W.; Hunt, W.F.; Ashley, R.; Butler, D.; Arthur, S.; Trowsdale, S.; Barraud, S.; Semadeni-Davies, A.; Bertrand-Krajewski, J.-L.; et al. SUDS, LID, BMPs, WSUD and more—The Evolution and Application of Terminology Surrounding Urban Drainage. *Urban Water J.* **2015**, *12*, 525–542. [CrossRef]
5. Prince George's County Maryland. *Low-Impact Development Design Strategies an Integrated Design Approach Low-Impact Development: An Integrated Design Approach*; Department of Environmental Resources: Largo, MD, USA, 1999.

6. Cheng, M.-S.; Coffman, L.S.; Clar, M.L. Low-Impact Development Hydrologic Analysis. In *Urban Drainage Modeling: Proceedings of the Specialty Symposium of the World Water and Environmental Resources Congress, Orlando, FL, USA, 20-24 May 2001*; American Society of Civil Engineers: Reston, VA, USA, 2001; pp. 659–681.

7. Coffman, L.S. Low Impact Development: Smart Technology for Clean Water. In Proceedings of the Ninth International Conference on Urban Drainage, Portland, OR, USA, 8–13 September 2002; American Society for Civil Engineers: Reston, VA, USA, 2002; pp. 1–11.

8. Khan, U.T.; Valeo, C.; Chu, A.; He, J. A Data Driven Approach to Bioretention Cell Performance: Prediction and Design. *Water* **2013**, *5*, 13–28. [CrossRef]

9. Khan, U.T.; Valeo, C.; Chu, A.; van Duin, B. Bioretention Cell Efficacy In Cold Climates: Part 1—Hydrologic Performance. *Can. J. Civ. Eng.* **2012**, *39*, 1210–1221. [CrossRef]

10. Khan, U.T.; Valeo, C.; Chu, A.; van Duin, B. Bioretention Cell Efficacy In Cold Climates: Part 2—Water Quality Performance. *Can. J. Civ. Eng.* **2012**, *39*, 1222–1233. [CrossRef]

11. Ishaq, S.; Hewage, K.; Farooq, S.; Sadiq, R. State of provincial regulations and guidelines to promote low impact development (LID) alternatives across Canada: Content analysis and comparative assessment. *J. Environ. Manag.* **2019**, *235*, 389–402. [CrossRef]

12. Elliott, A.H.; Trowsdale, S.A. A Review of Models for Low Impact Urban Stormwater Drainage. *Environ. Model. Softw.* **2007**, *22*, 394–405. [CrossRef]

13. Coffman, L.S.; Goo, R.; Frederick, R. Low-Impact Development An Innovative Alternative Approach to Stormwater Management. In Proceedings of the 29th Annual Water Resources Planning and Management Conference, Tempe, AZ, USA, 6–9 June 1999; pp. 1–10.

14. Vogel, J.R.; Moore, T.L.; Coffman, R.R.; Rodie, S.N.; Hutchinson, S.L.; McDonough, K.R.; McLemore, A.J.; McMaine, J.T. Critical Review of Technical Questions Facing Low Impact Development and Green Infrastructure: A Perspective from the Great Plains. *Water Environ. Res.* **2015**, *87*, 849–862. [CrossRef]

15. Johns, C.; Shaheen, F.; Woodhouse, M. *Green Infrastructure and Stormwater Management in Toronto: Policy Context and Instruments, Centre for Urban Research and Land Development*; Centre for Urban Research and Land Development: Toronto, ON, Canada, 2018.

16. Li, J.; Deng, C.; Li, Y.; Li, Y.; Song, J. Comprehensive Benefit Evaluation System for Low-Impact Development of Urban Stormwater Management Measures. *Water Resour. Manag.* **2017**, *31*, 4745–4758. [CrossRef]

17. Mao, X.; Jia, H.; Yu, S.L. Assessing the ecological benefits of aggregate LID-BMPs through modelling. *Ecol. Modell.* **2017**, *353*, 139–149. [CrossRef]

18. Schifman, L.A.; Prues, A.; Gilkey, K.; Shuster, W.D. Realizing the opportunities of black carbon in urban soils: Implications for water quality management with green infrastructure. *Sci. Total Environ.* **2018**, *644*, 1027–1035. [CrossRef] [PubMed]

19. Xie, N.; Akin, M.; Shi, X. Permeable concrete pavements: A review of environmental benefits and durability. *J. Clean. Prod.* **2019**, *210*, 1605–1621. [CrossRef]

20. Seo, M.; Jaber, F.; Srinivasan, R.; Jeong, J. Evaluating the Impact of Low Impact Development (LID) Practices on Water Quantity and Quality under Different Development Designs Using SWAT. *Water* **2017**, *9*, 193. [CrossRef]

21. Capotorti, G.; Alós Ortí, M.M.; Copiz, R.; Fusaro, L.; Mollo, B.; Salvatori, E.; Zavattero, L. Biodiversity and ecosystem services in urban green infrastructure planning: A case study from the metropolitan area of Rome (Italy). *Urban For. Urban Green.* **2019**, *37*, 87–96. [CrossRef]

22. Morakinyo, T.E.; Lam, Y.F.; Hao, S. Evaluating the role of green infrastructures on near-road pollutant dispersion and removal: Modelling and measurement. *J. Environ. Manag.* **2016**, *182*, 595–605. [CrossRef]

23. Nordbo, A.; Järvi, L.; Haapanala, S.; Wood, C.R.; Vesala, T. Fraction of natural area as main predictor of net CO_2 emissions from cities. *Geophys. Res. Lett.* **2012**, *39*. [CrossRef]

24. Chenoweth, J.; Anderson, A.R.; Kumar, P.; Hunt, W.F.; Chimbwandira, S.J.; Moore, T.L.C. The interrelationship of green infrastructure and natural capital. *Land Use Policy* **2018**, *75*, 137–144. [CrossRef]

25. Rafael, S.; Vicente, B.; Rodrigues, V.; Miranda, A.I.; Borrego, C.; Lopes, M. Impacts of green infrastructures on aerodynamic flow and air quality in Porto's urban area. *Atmos. Environ.* **2018**, *190*, 317–330. [CrossRef]

26. Jayasooriya, V.M.; Ng, A.W.M.; Muthukumaran, S.; Perera, B.J.C. Urban Forestry & Urban Greening Green infrastructure practices for improvement of urban air quality. *Urban For. Urban Green.* **2017**, *21*, 34–47.

27. Moore, T.L.C.; Hunt, W.F. Ecosystem service provision by stormwater wetlands and ponds—A means for evaluation? *Water Res.* **2012**, *46*, 6811–6823. [CrossRef] [PubMed]

28. Hassall, C.; Anderson, S. Stormwater ponds can contain comparable biodiversity to unmanaged wetlands in urban areas. *Hydrobiologia* **2015**, *745*, 137–149. [CrossRef]

29. Hassall, C. The ecology and biodiversity of urban ponds. *Wiley Interdiscip. Rev. Water* **2014**, *1*, 187–206. [CrossRef]

30. Taylor, J.R.; Lovell, S.T. Urban home food gardens in the Global North: Research traditions and future directions. *Agric. Hum. Values* **2014**, *31*, 285–305. [CrossRef]

31. Hostetler, M.; Allen, W.; Meurk, C. Landscape and Urban Planning Conserving urban biodiversity? Creating green infrastructure is only the first step. *Landsc. Urban Plan.* **2011**, *100*, 369–371. [CrossRef]

32. Pinho, P.; Correia, O.; Lecoq, M.; Munzi, S.; Vasconcelos, S.; Gonçalves, P.; Rebelo, R.; Antunes, C.; Silva, P.; Freitas, C.; et al. Evaluating green infrastructure in urban environments using a multi-taxa and functional diversity approach. *Environ. Res.* **2016**, *147*, 601–610. [CrossRef]

33. Ozturk, Z.C.; Dursun, S. Low Impact Development and Green Infrastructure. *Sci. J. Environ. Sci.* **2016**, *5*, 75–80.

34. Winz, I.; Brierley, G.; Trowsdale, S. Dominant perspectives and the shape of urban stormwater futures. *Urban Water J.* **2011**, *8*, 337–349. [CrossRef]

35. Russo, A.; Escobedo, F.J.; Cirella, G.T.; Zerbe, S. Edible green infrastructure: An approach and review of provisioning ecosystem services and disservices in urban environments. *Agric. Ecosyst. Environ.* **2017**, *242*, 53–66. [CrossRef]

36. Kevern, J.T.; Asce, A.M. Green Building and Sustainable Infrastructure: Sustainability Education for Civil Engineers. *J. Prof. Issues Eng. Educ. Pract.* **2011**, *137*, 107–112. [CrossRef]

37. Han, H.; Li, H.; Zhang, K. Urban Water Ecosystem Health Evaluation Based on the Improved Fuzzy Matter-Element Extension Assessment Model: Case Study from Zhengzhou City, China. *Math. Probl. Eng.* **2019**, *2019*, 7502342. [CrossRef]

38. Markevych, I.; Schoierer, J.; Hartig, T.; Chudnovsky, A.; Hystad, P.; Dzhambov, A.M.; de Vries, S.; Triguero-Mas, M.; Brauer, M.; Nieuwenhuijsen, M.J.; et al. Exploring pathways linking greenspace to health: Theoretical and methodological guidance. *Environ. Res.* **2017**, *158*, 301–317. [CrossRef] [PubMed]

39. Wood, L.; Hooper, P.; Foster, S.; Bull, F. Public green spaces and positive mental health—Investigating the relationship between access, quantity and types of parks and mental wellbeing. *Health Place* **2017**, *48*, 63–71. [CrossRef] [PubMed]

40. United States Environmental Protection Agency (EPA) Terminology of Low Impact Development: Distinguishing LID from Other Techniques That Address Community Growth Issues. Available online: https://www.epa.gov/sites/production/files/2015-09/documents/bbfs2terms.pdf (accessed on 6 November 2019).

41. Eckart, K.; McPhee, Z.; Bolisetti, T. Performance and implementation of low impact development—A review. *Sci. Total Environ.* **2017**, *607–608*, 413–432. [CrossRef]

42. Kändler, N.; Annus, I.; Vassiljev, A.; Puust, R.; Kaur, K. Smart In-Line Storage Facilities in Urban Drainage Network. *Proceedings* **2018**, *2*, 631.

43. Pochwat, K.; Iličić, K. A simplified dimensioning method for high-efficiency retention tanks. E3S Web Conf. VI International Conference of Science and Technology INFRAEKO 2018 Modern Cities. *Infrastruct. Environ.* **2018**, *45*, 1–8.

44. Chang, C.L.; Lo, S.L.; Huang, S.M. Optimal strategies for best management practice placement in a synthetic watershed. *Environ. Monit. Assess.* **2009**, *153*, 359–364. [CrossRef]

45. Ariza, S.L.J.; Martínez, J.A.; Muñoz, A.F.; Quijano, J.P.; Rodríguez, J.P.; Camacho, L.A.; Díaz-Granados, M. A Multicriteria Planning Framework to Locate and Select Sustainable Urban Drainage Systems (SUDS) in Consolidated Urban Areas. *Sustainability* **2019**, *11*, 2312. [CrossRef]

46. Kuller, M.; Bach, P.M.; Roberts, S.; Browne, D.; Deletic, A. A planning-support tool for spatial suitability assessment of green urban stormwater infrastructure. *Sci. Total Environ.* **2019**, *686*, 856–868. [CrossRef]

47. Kuller, M.; Bach, P.M.; Ramirez-Lovering, D.; Deletic, A. Framing water sensitive urban design as part of the urban form: A critical review of tools for best planning practice. *Environ. Model. Softw.* **2017**, *96*, 265–282. [CrossRef]

48. Kaykhosravi, S.; Khan, U.; Jadidi, A.; Kaykhosravi, S.; Khan, U.T.; Jadidi, A. A Comprehensive Review of Low Impact Development Models for Research, Conceptual, Preliminary and Detailed Design Applications. *Water* **2018**, *10*, 1541. [CrossRef]

49. Malczewski, J.; Rinner, C. *Multicriteria Decision Analysis in Geographic Information Science*; Springer: New York, NY, USA, 2015; ISBN 9783540747567.

50. Zischg, J.; Zeisl, P.; Winkler, D.; Rauch, W.; Sitzenfrei, R. On the sensitivity of geospatial low impact development locations to the centralized sewer network. *Water Sci. Technol.* **2018**, *77*, 1851–1860. [CrossRef] [PubMed]

51. Bach, P.M.; McCarthy, D.T.; Urich, C.; Sitzenfrei, R.; Kleidorfer, M.; Rauch, W.; Deletic, A. A planning algorithm for quantifying decentralised water management opportunities in urban environments. *Water Sci. Technol.* **2013**, *68*, 1857–1865. [CrossRef] [PubMed]

52. Song, J.Y.; Chung, E. A Multi-Criteria Decision Analysis System for Prioritizing Sites and Types of Low Impact Development Practices: Case of Korea. *Water* **2017**, *9*, 291. [CrossRef]

53. Ahmed, K.; Chung, E.-S.; Song, J.-Y.; Shahid, S. Effective Design and Planning Specification of Low Impact Development Practices Using Water Management Analysis Module (WMAM): Case of Malaysia. *Water* **2017**, *9*, 173. [CrossRef]

54. Chung, E.S.; Hong, W.P.; Lee, K.S.; Burian, S.J. Integrated Use of a Continuous Simulation Model and Multi-Attribute Decision-Making for Ranking Urban Watershed Management Alternatives. *Water Resour. Manag.* **2010**, *25*, 641–659. [CrossRef]

55. Lee, J.G.; Selvakumar, A.; Alvi, K.; Riverson, J.; Zhen, J.X.; Shoemaker, L.; Lai, F. A watershed-scale design optimization model for stormwater best management practices. *Environ. Model. Softw.* **2012**, *37*, 6–18. [CrossRef]

56. Charlesworth, S.; Warwick, F.; Lashford, C. Decision-Making and Sustainable Drainage: Design and Scale. *Sustainability* **2016**, *8*, 782. [CrossRef]

57. Jato-Espino, D.; Sillanpää, N.; Charlesworth, S.M.; Andrés-Doménech, I. Coupling GIS with Stormwater Modelling for the Location Prioritization and Hydrological Simulation of Permeable Pavements in Urban Catchments. *Water* **2016**, *8*, 451. [CrossRef]

58. Yang, J.S.; Son, M.W.; Chung, E.S.; Kim, I.H. Prioritizing Feasible Locations for Permeable Pavement Using MODFLOW and Multi-criteria Decision Making Methods. *Water Resour. Manag.* **2015**, *29*, 4539–4555. [CrossRef]

59. Lerer, S.; Arnbjerg-Nielsen, K.; Mikkelsen, P. A Mapping of Tools for Informing Water Sensitive Urban Design Planning Decisions—Questions, Aspects and Context Sensitivity. *Water* **2015**, *7*, 993–1012. [CrossRef]

60. Martin-Miklea, C.J.; de Beurs, K.M.; Julianb, J.P.; Mayer, P.M. Identifying priority sites for low impact development (LID) in a mixed-use watershed. *Landsc. Urban Plan.* **2015**, *140*, 29–41. [CrossRef]

61. Walter, M.T.; Walter, M.F.; Brooks, E.S.; Steenhuis, T.S.; Boll, J.; Weiler, K. Hydrologically Sensitive Areas: Variable Source Area Hydrology Implications for Water Quality Risk Assessment. *J. Soil Water Conserv.* **2000**, *55*, 277–284.

62. Zhang, A.; Shi, H.; Li, T.; Fu, X. Analysis of the influence of rainfall spatial uncertainty on hydrological simulations using the bootstrap method. *Atmosphere* **2018**, *9*, 71. [CrossRef]

63. Natural Resources Conservation Service. *National Engineering Handbook Chapter 4 Storm Rainfall Depth*; United States Department of Agriculture: Washington DC, USA, 2015.

64. Zhang, K.; Chui, T.F.M. A Comprehensive Review of Spatial Allocation of LID-BMP-GI Practices: Strategies and Optimization Tools. *Sci. Total Environ.* **2018**, *621*, 915–929. [CrossRef]

65. Bach, P.M.; Rauch, W.; Mikkelsen, P.S.; McCarthy, D.T.; Deletic, A. A critical review of integrated urban water modelling—Urban drainage and beyond. *Environ. Model. Softw.* **2014**, *54*, 88–107. [CrossRef]

66. Nocco, M.A.; Rouse, S.E.; Balster, N.J. Vegetation type alters water and nitrogen budgets in a controlled, replicated experiment on residential-sized rain gardens planted with prairie, shrub, and turfgrass. *Urban Ecosyst.* **2016**, *19*, 1665–1691. [CrossRef]

67. Toronto and Region Conservation Authority (TRCA) Low Impact Development. Available online: https://trca.ca/conservation/restoration/low-impact-development/ (accessed on 6 November 2019).

68. Mein, R.G.; Larson, C.L. Modeling infiltration during a steady rain. *Water Resour. Res.* **1973**, *9*, 384–394. [CrossRef]

69. McCuen, R.H.; Wong, S.L.; Rawls, W.J. Estimating urban time of concentration. *J. Hydraul. Eng.* **1984**, *110*, 887–904. [CrossRef]

70. Chow, V.T.; Maidment, D.R.; Mays, L.W. *Applied Hydrology*; McGrawHill: New York, NY, USA, 1988.

71. European Environment Agency (EEA). *Environmental Indicators: Typology and Overview*; European Environment Agency: Copenhagen, Denmark, 1999.

72. Matos, P.; Vieira, J.; Rocha, B.; Branquinho, C.; Pinho, P. Modeling the provision of air-quality regulation ecosystem service provided by urban green spaces using lichens as ecological indicators. *Sci. Total Environ.* **2019**, *665*, 521–530. [CrossRef]

73. Buccolieri, R.; Jeanjean, A.P.R.; Gatto, E.; Leigh, R.J. The impact of trees on street ventilation, NOx and $PM_{2.5}$ concentrations across heights in Marylebone Rd street canyon, central London. *Sustain. Cities Soc.* **2018**, *41*, 227–241. [CrossRef]

74. Kyrö, K.; Brenneisen, S.; Kotze, D.J.; Szallies, A.; Gerner, M.; Lehvävirta, S. Local habitat characteristics have a stronger effect than the surrounding urban landscape on beetle communities on green roofs. *Urban For. Urban Green.* **2018**, *29*, 122–130. [CrossRef]

75. Pei, N.; Wang, C.; Jin, J.; Jia, B.; Chen, B.; Qie, G.; Qiu, E.; Gu, L.; Sun, R.; Li, J.; et al. Long-term afforestation efforts increase bird species diversity in Beijing, China. *Urban For. Urban Green.* **2018**, *29*, 88–95. [CrossRef]

76. Lucke, T.; Nichols, P.W.B. The pollution removal and stormwater reduction performance of street-side bioretention basins after ten years in operation. *Sci. Total Environ.* **2015**, *536*, 784–792. [CrossRef] [PubMed]

77. Kluge, B.; Markert, A.; Facklam, M.; Sommer, H.; Kaiser, M.; Pallasch, M.; Wessolek, G. Metal accumulation and hydraulic performance of bioretention systems after long-term operation. *J. Soils Sediments* **2018**, *18*, 431–441. [CrossRef]

78. Ma, Y.; He, W.; Zhao, H.; Zhao, J.; Wu, X.; Wu, W.; Li, X.; Yin, C. Influence of Low Impact Development practices on urban diffuse pollutant transport process at catchment scale. *J. Clean. Prod.* **2019**, *213*, 357–364. [CrossRef]

79. Passeport, E.; Vidon, P.; Forshay, K.J.; Harris, L.; Kaushal, S.S.; Kellogg, D.Q.; Lazar, J.; Mayer, P.; Stander, E.K. Ecological Engineering Practices for the Reduction of Excess Nitrogen in Human-Influenced Landscapes: A Guide for Watershed Managers. *Environ. Manag.* **2013**, *51*, 392–413. [CrossRef]

80. Weitman, D.; Weinberg, A.; Goo, R. Reducing stormwater costs through LID strategies and practices. In Proceedings of the Low Impact Development for Urban Ecosystem and Habitat Protection, Seattle, WA, USA, 16–19 November 2008.

81. Shuster, W.D.; Gehring, R.; Gerken, J. Prospects for enhanced groundwater recharge via infiltration of urban storm water runoff: A case study. *J. Soil Water Conserv.* **2007**, *62*, 129–137.

82. Deeb, M.; Groffman, P.M.; Joyner, J.L.; Lozefski, G.; Paltseva, A.; Lin, B.; Mania, K.; Cao, D.L.; McLaughlin, J.; Muth, T.; et al. Soil and microbial properties of green infrastructure stormwater management systems. *Ecol. Eng.* **2018**, *125*, 68–75. [CrossRef]

83. McPhillips, L.; Walter, M.T. Hydrologic conditions drive denitrification and greenhouse gas emissions in stormwater detention basins. *Ecol. Eng.* **2015**, *85*, 67–75. [CrossRef]

84. Hunt, W.F.; Davis, A.P.; Traver, R.G. Meeting Hydrologic and Water Quality Goals Through Targeted Bioretention Design. *J. Environ. Eng.* **2012**, *138*, 698–707. [CrossRef]

85. Ballard, B.W.; Wilson, S.; Udale-Clarke, H.; Illman, S.; Scott, T.; Ashley, R.; Kellagher, R. *The SUDS Manual*; Constuction Industry Research and Informsation Association: London, UK, 2015.

86. Meerow, S.; Newell, J.P. Spatial planning for multifunctional green infrastructure: Growing resilience in Detroit. *Landsc. Urban Plan.* **2017**, *159*, 62–75. [CrossRef]

87. Shackleton, C.M.; Blair, A.; De Lacy, P.; Kaoma, H.; Mugwagwa, N.; Dalu, M.T.; Walton, W. How important is green infrastructure in small and medium-sized towns? Lessons from South Africa. *Landsc. Urban Plan.* **2018**, *180*, 273–281. [CrossRef]

88. Kardan, O.; Gozdyra, P.; Misic, B.; Moola, F.; Palmer, L.J.; Paus, T.; Berman, M.G. Neighborhood greenspace and health in a large urban center. *Sci. Rep.* **2015**, *5*, 11610. [CrossRef] [PubMed]

89. Dadvand, P.; Poursafa, P.; Heshmat, R.; Motlagh, M.E.; Qorbani, M.; Basagaña, X.; Kelishadi, R. Use of green spaces and blood glucose in children; a population-based CASPIAN-V study. *Environ. Pollut.* **2018**, *243*, 1134–1140. [CrossRef]

90. Kuo, M. How might contact with nature promote human health? Promising mechanisms and a possible central pathway. *Front. Psychol.* **2015**, *6*, 1093. [CrossRef]

91. Scott, J.T.; Kilmer, R.P.; Wang, C.; Cook, J.R.; Haber, M.G. Natural environmental near schools: Potential benefits for socio-emotional and behavioural development in early childhood. *Am. J. Community Psychol.* **2018**, *62*, 419–432. [CrossRef]

92. Leung, W.T.V.; Yee Tiffany Tam, T.; Pan, W.C.; Wu, C.-D.; Lung, S.-C.C.; Spengler, J.D. How is environmental greenness related to students' academic performance in English and Mathematics? *Landsc. Urban Plan.* **2019**, *181*, 118–124. [CrossRef]

93. Liu, G.C.; Wilson, J.S.; Qi, R.; Ying, J. Green neighborhoods, food retail and childhood overweight: Differences by population density. *Am. J. Health Promot.* **2007**, *21*, 317–325. [CrossRef]

94. Kwak, D.; Kim, H.; Han, M. Runoff Control Potential for Design Types of Low Impact Development in Small Developing Area Using XPSWMM. *Procedia Eng.* **2016**, *154*, 1324–1332. [CrossRef]

95. Kamenetzky, R.D. The Relationship Between the Analytic Hierarchy Process and the Additive Value Function. *Decis. Sci.* **1982**, *13*, 702–713. [CrossRef]

96. Pereira, J.M.C.; Duckstein, L. A multiple criteria decision-making approach to gis-based land suitability evaluation. *Int. J. Geogr. Inf. Syst.* **1993**, *7*, 407–424. [CrossRef]

97. U.S. Army Corps of Engineers Hydrological Engineer Center. Available online: http://www.hec.usace.army.mil/software/hec-hms/features.aspx (accessed on 6 November 2019).

98. Keller, G.; Sherar, J. *Low-volume roads engineering: Best management practices field guide*; United States Forest Service: Washington, DC, USA, 2005; Chapter 11; pp. 103–114.

99. Muir, R.J. Flood Insurance in Canada. Available online: https://www.cityfloodmap.com/2013/12/flood-insurance-in-canada.html (accessed on 6 November 2019).

Article

Climate and Land Use Influences on Bacteria Levels in Stormwater

Kaifeng Xu [1], Caterina Valeo [1,*], Jianxun He [2] and Zhiying Xu [1]

[1] Department of Mechanical Engineering, University of Victoria, Victoria, BC V8W 2Y2, Canada; xukaifengian@gmail.com (K.X.); sunny520leo@gmail.com (Z.X.)

[2] Civil Engineering, Schulich School of Engineering, University of Calgary, Calgary, AB T2N 1N4, Canada; jianhe@ucalgary.ca

* Correspondence: valeo@uvic.ca; Tel.: +1-250-721-8623

Received: 31 October 2019; Accepted: 18 November 2019; Published: 22 November 2019

Abstract: The influence of climatic variables and land use on fecal coliform (FC) levels in stormwater collected from outfalls throughout southern Vancouver Island between 1995 and 2011 are examined through statistical analyses, Fourier analysis, Multiple Linear Regression (LR) and Multivariate Logistic Regression (MLR). Kendall's τ-b demonstrated that FC levels were significantly and positively correlated with the amount of residential area within a drainage catchment generating the runoff, and that FC levels were location dependent. Climatic variables of temperature and antecedent dry period length were significantly and positively correlated with FC levels at both the sampling location level and across the region overall. Precipitation and flowrates were negatively correlated with FC levels. Fourier analysis showed that monthly FC levels shared the same 12 month cycle (peaking in July) as precipitation and temperature. MLR modelling was applied by aggregating the LogFC data by order of magnitude. The MLR model shows that the data are subject to different influences depending on the season and as well, the month of the year. The land use and climate analyses suggest that future climate change impact studies attempted on nearshore bacterial water quality should be conducted at the urban catchment scale.

Keywords: stormwater quality; fecal coliforms; Vancouver Island; nearshore areas; bacteria loading; multinomial logistic regression; periodicity analysis; land use impacts; climate impacts

---

1. Introduction

Contaminates transported through stormwater runoff to coastal waters can pose a potential risk to public health and the environment [1]. Furthermore, the stormwater drainage system can also be contaminated by sewage through infiltration or unintended connections with sewer systems and poorly maintained in-ground sewage disposal systems [2]. Contaminated stormwater runoff will often contain excessive levels of bacterial contaminants, which are directly related to disease outbreaks and adverse impacts to aquatic life [3,4]. Therefore, many municipalities will monitor and manage stormwater quality for both health and environmental concerns. Fecal coliforms (FC) have historically been used as a fecal indicator to indicate the presence of microbial contamination in surface and ground waters [5,6]. Since they are considered as an indicator of surface water quality and safety, FC is often selected for monitoring in those water quality monitoring programs concerned with microbial loading. Microbially contaminated water can be a serious source of intestinal disease through ingestion, or exposure through bathing or by consuming contaminated shellfish [7]. Factors that could affect fecal indicator bacteria levels have been investigated in the literature [8–10]. Sewage overflow, wildlife and stormwater runoff from urban and agricultural land use are important sources of fecal coliform affecting water quality [11,12]. Non-point urban and agricultural land use zones have significant impacts on water quality and produce a large number of fecal bacteria conveyed to water bodies [13,14].

As well, climatic variables like precipitation and wet seasons in a region are suggested to have a positive correlation with the concentration of fecal bacteria in the surface waters [15–17].

There are a variety of studies looking at stormwater generated microbial loadings and how they are influenced by climatic variables. Henry et al. [18] found rainfall within the 24 h preceding sampling was found to be correlated with incidence of fecal matter, however, absence of rainfall was also significantly tied the incidence of fecal matter depending on the location. The length of the period over which precipitation was computed for determining correlations with fecal contamination, impacted whether the correlations were positive, negative, significant, or had no impact [19–23]. Seasonal variations in the correlations computed for fecal indicators versus temperature, precipitation and antecedent dry period length were observed in several studies [21,24,25] but generally, temperature was positively correlated with indicator bacteria concentrations [26], and precipitation was also positively correlated [20,22]. McCarthy et al. [27] found that *E. coli* (EC) levels were highly correlated to antecedent climatic parameters but like [26], found they were less correlated to hydrologic parameters such as runoff and suggested that EC concentrations "were not prone to displaying a first flush effect", which is in contrast to other studies in which fecal coliforms peaked after storm events [28]. Solar intensity was also seen to have a negative correlation with EC [29]. The additional influence of spatial location in analyses examining climatic influences on fecal contamination was observed in several studies [30,31] with Vermeulen and Hofstra [31] suggesting that the great variability among location characteristics helped to explain variations in *E. coli* levels in their study. All of these studies vary extensively by how strongly or weakly correlations are seen between bacteria loading and climate variables, as well as to the way the variable is constructed. That is, depending on whether precipitation is measured during a rain event, on the day of sampling, or averaged over a 3 day period over which the sampling took place, produced variations in results. This suggests that the scale of the climate variable matters as much in the research as the climatic variable itself (whether temperature, precipitation, solar radiation, etc.).

Spatial influence studies in the literature include several works stating that the contamination in coastal water quality is caused by the combined effects of human activity and environmental factors in coastal areas [28,32]. The accumulated fecal coliform can be transported into nearshore ocean regions from direct runoff or sewage overflow during storm events [33,34] in combined sewer systems. Mallin et al. [35] showed that watershed population and watershed size were significantly related to average fecal coliform levels, with the strongest relationships for percentage of impervious surface and fecal coliform levels. Jent et al [19] found that in non-dry periods, fecal contamination was significantly correlated with agricultural land use, while Tiefenthaler et al. [21] showed that mean EC were significantly greater in developed watersheds than undeveloped watersheds. Similarly, Vitro et al. [36] found that road network density was associated with increasing fecal coliform levels but housing unit density had a significantly negative relationship with FC levels; Paule-Mercado et al. [25] found the highest FC concentrations were also found in urban areas versus agricultural areas. Delpla and Rodriguez [37] showed that FC were significantly and positively correlated with urban and agricultural land, but forests were negatively correlated. FC concentration was found to be significantly and positively correlated with urban land use in even low percentages of urbanization [38]. Additional studies [20,22] found higher fecal indicator bacteria concentrations in urban sites over forested sites. The literature suggests that urban areas provide higher bacterial loads in comparison to undeveloped green spaces, and agricultural lands also contribute higher bacterial loads than green spaces.

Other studies have gone a step further in attempts to mathematically model bacteria levels as a function of land-use or climatic variables in order to provide insight into the relationships. Wu et al. [23] used logistic regression modelling and showed that EC *presence* was significantly and positively associated with developed area but negatively associated with agricultural area. Multivariable logistic regression models were used to show that increases in the number of heavy rain days significantly increased the likelihood of the presence of EC and that EC was also likely to be detected with increases in mean temperature. Cha et al. [24] developed a model using Bayesian regression that used meteorological and land-use characteristics to estimate FC concentrations. The modelling was used to provide predictions in FC increases given temperature increases. Galfi et al. [39] used PCA and cluster analysis to find correlation patterns and found that significant climatic variables for influencing indicator bacteria levels varied among catchments as well as seasonally. St. Laurent and Mazumder [38] used a classification tree to better understand the influence of land use on FC levels. A common aspect of all of these modelling studies is that they use data-driven, stochastic methods for modelling. This is reasonable as causal mechanisms are difficult to capture and physically based models may be onerous to implement when simply attempting to determine climatic influences on FC concentrations. Examples of sophisticated, physically-based models in the literature that predict bacterial levels as a function of environmental parameters include [29,40], but both look at transport in river systems. The latter did note that the sparsity in the time series of bacteria concentrations adversely affected the modelling [40].

The Capital Regional District (CRD) of southern Vancouver Island is a government body representing 13 municipalities and three electoral areas. The core area of CRD has approximately 270,000 people and 27,600 hectares including residential, industrial, commercial, institutional and agriculture regions. The area has approximately 8350 properties with onsite sewage disposal and with others connected to a sewer system [2]. The CRD assists these municipalities in developing their stormwater management plans and infrastructure, as well as water quality monitoring for these municipalities. Coastal water quality has been a critical issue worldwide for nearshore inhabited regions and the CRD is no exception. Coastal water quality is known to be affected by natural and human factors includes runoff, sewage wastewater, land reclamation and climate change [41]. The implementation of monitoring programs is necessary and critical in managing coastal water contamination [42,43]. The CRD collects pollutant levels (including fecal coliforms) within stormwater pipes, streams and nearshore areas throughout the CRD, with an interest in identifying hotspots and remediating those areas of highest priority. This process includes collecting water samples and analyzing for fecal coliform bacteria in the sample. The collected data are used to estimate and analyze the distribution of microbial contamination and any possible public health concerns. This allows the jurisdictions involved to undertake remedial measures where most needed. Currently, the sampling frequency is regulated primarily by cost and capacity but increased sampling strategies are advised for locations exceeding a certain threshold. Given that fecal coliform contamination in stormwater runoff is directly influenced by climate and watershed properties [10,13,44], a regular sampling scheme that does not consider climate and weather will likely miss peaks in contamination.

The literature shows some consensus on the impact of land use on bacteria levels but the influence of climate variables remains contentious, particularly where the study focuses on stormwater runoff generation. Much of what is observed is likely due to site specificity and the scale of the sampling and analysis [40,45]. If agricultural areas are absent from the drainage area, bacteria loads in stormwater runoff generation are believed to be low and are thus, highly affected by sampling plans, which tend to be sparse and intermittent. Thus, stormwater specific data sets tend to be limited in terms of sampling period, length of database in time and extent over space, and therefore, the insight they can provide on environmental influences is limited. This often drives researchers to turn to data-driven methods [23,24,38,39,46,47] to provide insight but these methods are not unaffected by the temporal and spatial scale of the sampling and analysis.

Given the extensive monitoring network developed by the CRD, the objectives of this work are to determine if the fecal coliform data collected by the CRD during its regular monitoring program are influenced by local weather and land use; what insights arise from the scale of the collection and variable (fecal coliforms in this research) studied, and in particular, whether data-driven methods are able to provide insights into the causal mechanisms behind the observations. With the outcomes, the monitoring program can be modified to improve sampling statistics and possibly indicate the impacts of climate change on bacterial loads in the future.

2. Materials and Methods

2.1. Study Area

The study area is located on the southeastern core area of the CRD between latitudes 48°29′57.3″ N and 48°24′02.2″ N, and longitudes 123°26′26.3″ W and 123°15′40.4″ W. Shown in Figure 1, the study area forms part of the "core area" of the CRD and is approximately 15,000 hectares in size and spanning six municipalities, two First Nations, federal, provincial, regional, and municipal parkland. Land use in the core includes residential, commercial, industrial, institutional, and agricultural activities and is shown in Figure 2b. The region lies within the *Coast Mountains and Islands* physiographic region, and contains a number of watersheds of variable size, some natural and some urbanized, draining to the coastline through stormwater drainage systems or creeks and rivers. Within the *Georgia Depression* Ecozone, the Biogeoclimatic zone is *Coastal Douglas Fir*, which experiences mean annual temperatures of almost 10 °C and mean annual precipitation of over 1000 mm annually [48]. The mean coldest month is January with average temperatures of around 3 °C and the mean warmest month is in July, which averages roughly 17 °C. This mild climate experiences on average 204 frost-free days per year. While weather systems in British Columbia are highly impacted by elevation changes, the elevation change in the study region is less than 50 m. Rainfall systems for this area tend to form in the north Pacific and move easterly across the mountain range running down the center of the island. The process leads to greater rainfall on the western side of the Island and less on the leeward side (eastern side) resulting in a rain shadow for the study area because it is on the eastern side of the Island [48]. Rainfall is higher in the winter months and leads to stormwater runoff draining through the natural portions of the watershed (see Figure 2a) or through the stormwater minor drainage system and pipe system to end up discharged to the coastal harbors shown in Figure 1. Soils in the overburden above rock formations in pervious areas are generally Distric Brunisol [48] soil type within the Luvisolic soils group, which are typically well draining.

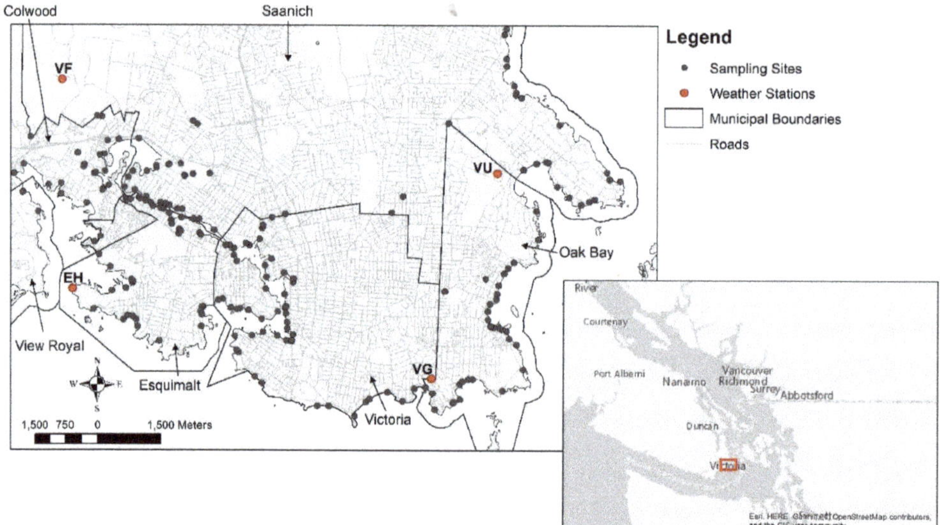

Figure 1. Location of sampling sites and meteorological stations along Nearshore Areas of Southern Vancouver Island. Municipal boundaries are marked with a solid dark line.

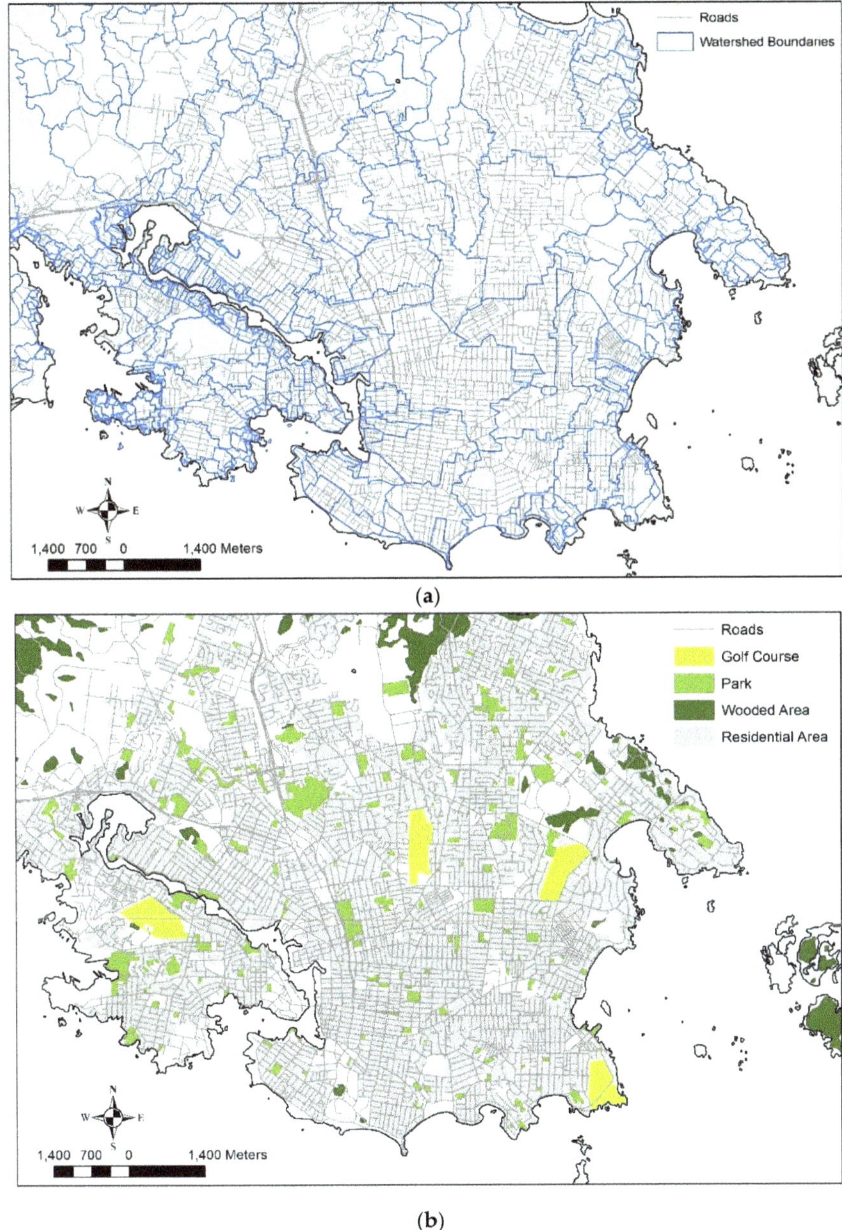

Figure 2. (**a**) Watershed map in Capital Region District; (**b**) Land-use map in Capital Region District.

2.2. Database Construction

The fecal coliform samples were collected from stormwater outfalls along the coastline of Esquimalt, Victoria, Oak Bay, and Saanich, and analyzed by the Stormwater, Harbours and Watersheds Program (SHWP) from 1995 to 2011. The database contains in total 6112 sampling data (N = 6112) from over 480 unique sampling locations. Stormwater samples for most stations were collected from each discharge point once during January to April (wet season) and once during June to September (dry season); although, there are data in all months of the year for most years in the database. Stormwater flows were sampled by land or boat at the point of discharge before going into the ocean to avoid unwanted flows. The sampling process attempts to avoid first flush conditions in order to reduce the chances of an unusual result. The measurements are also compared to historical results where the discharge is flagged for resampling outside the usual sampling period in the case of an unusually high observation. Thus for some years, there are data points for all 12 months [2].

A series of statistical tests were conducted on the data to determine which methods should be used in the analysis. The Kolmogorov–Smirnov test for normality showed that the fecal coliform data are not normally distributed and are highly right-skewed. In order to reduce the skewness and kurtosis, the fecal coliform (*FC*) data were log base10 transformed but the Kolmogorov–Smirnov test again rejected a normal distribution assumption for Log*FC*. All further analyses are conducted on *FC* data that are above a recreational water quality guideline of 200 CFU/100 mL [49] because data lower than this limit are not of concern in the environment. Restricting the database to values equal to, and above this limit resulted in a database of 2749 individual data observations over the 17 year period.

Contamination by sewage is considered an important factor causing high fecal coliform levels in waterbodies in many previous studies [11,12]. The report from SHWP also comments on the presence of sewage odors when a sample is taken, or other related problems. As a large amount of fecal bacteria exist in sewage wastewater, sewage cross-connections or infiltration can cause seriously adverse impacts. Therefore, non-parametric ANOVA is used to investigate if a significant difference exists in fecal coliform levels between the "sewage odor" identified stormwater samples and no odor stormwater samples. Sewage presence analysis was performed using the Kruskal Wallis Test and showed that the group of sewage-related samples (from odor detection) were significantly different from the group of regular samples with *p*-value = 0.000; the sewage-related samples have much higher levels of FC. There were 384 sewage-related samples. Since the focus of this study is stormwater that is not impacted by sewage, these data were removed from the final database used in the analysis. The frequency histogram of the final database is shown in Figure 3.

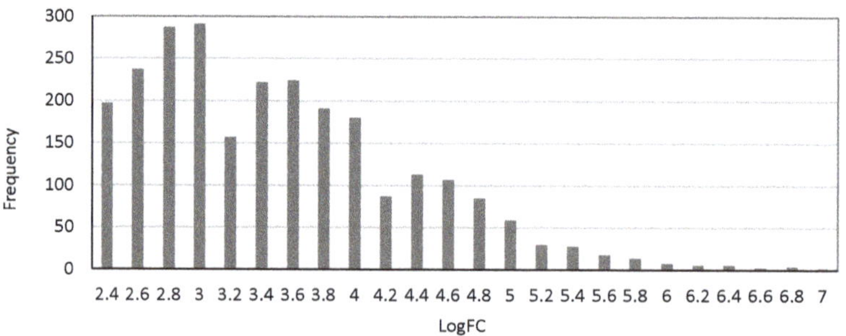

Figure 3. Histogram of cleaned Log*FC* data. The bin value indicates upper limit of that bin.

Daily historical data for this study were obtained from Environment Canada. The four weather stations selected in the CRD area, and shown in Figure 1, include the Esquimalt Harbour weather station, Victoria University CS weather station, Victoria Gonzales CS weather station and Victoria Francis Park weather station. Daily temperature and precipitation measured from 1995 to 2011 are used to check the relationship between bacterial contamination and meteorological variables. An average of the data from the four stations was computed and used in the analyses. Monthly climatic data averaged over the 17 years of collection were also generated for use in the periodicity analysis. In addition, cloud cover ratio data for the Victoria area were derived from Environment Canada, 1981–2010 station data for the study area. The watershed information shown in Figure 2a was obtained from the CRD and the land use information (Figure 2b) was obtained from Natural Resources Canada.

ArcGIS was used to visually explore the logarithm of the geometric mean of fecal coliform data above the regulated threshold of 200 CFU/100 mL collected within and adjacent to nearshore areas in the study region. The spatial distribution of bacterial contamination is visualized using the Kernel Density [50] which calculates the density of points in a region for each cell multiplied by the unit area. This map can provide a visual summary of the spatial distribution of fecal coliform data across the region as well as suggest any "hot spots". Figure 4 shows the areas determined as hot spots with high fecal coliform levels. The areas with deeper shading, indicating higher levels of FC are seen near the Victoria downtown, Esquimalt and the eastern shore area of Oak Bay. The southern shore of Oak Bay and the northeastern shores of Victoria have relatively lower pollution levels. The density map was also used to select hot-spot stations for further analysis (also shown in Figure 4). The five stations shown were selected to span the visually obvious areas of high FC levels, but also were stations that had consistent data collection over the period. The hot spots are located in the densely populated residential areas, which suggest that the source of fecal coliform is likely from urban activities. These five stations are further studied in the temporal analysis section.

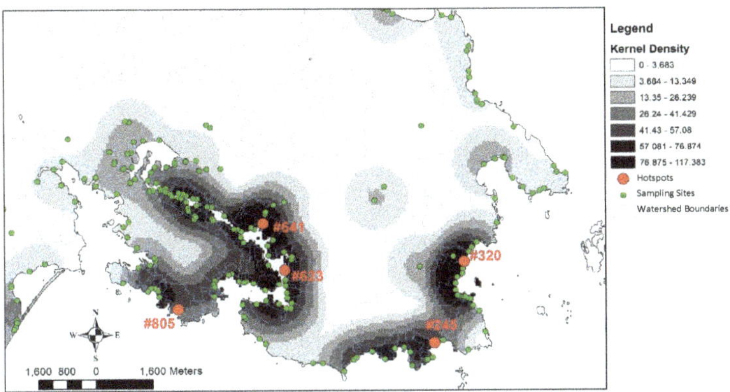

Figure 4. Density map based on the Log*FC* value of samples. Density values shown in the legend are in units of km^2.

For determining relationships between these data and climatic variables, correlations are computed between the data and several independent variables: 7, 3, 2 and 1-day rainfall totals preceding, or on the day of sampling; 7, 3, 2 and 1-day mean temperature; maximum temperature and minimum temperature preceding, or on the day of sampling; mean monthly cloud cover ratio; antecedent dry period length preceding the day of sampling, and flowrate measured at or close to the sampling location. Spatially distributed variables considered are watershed drainage area leading to the sampling point and land use in that drainage area, which is divided into residential use and green spaces.

2.3. Data Exploration and Analysis Methods

IBM SPSS, MATLAB and MS Excel are used in the statistical analyses and exploration of the data. To investigate trends and periodicity, dependent variables such as Log*FC*$_{i,j}$, are distinguished here from independent variables (such as air temperature and precipitation), where i represents the date collected and j indicates the station number. Other dependent variables used in this research include, the monthly (m) average of the log-transformed fecal coliform values, Log*FC*$_{y,m}$ in each year y over the 17 year period for all stations combined; and the 17 year average by month, log-transformed fecal coliform values (Log*FC*$_m$) is also considered. The geometric mean of the *FC*$_{i,j}$ values in the averaging period are computed first, then log transformed.

2.3.1. Periodicity and Fourier Analysis

Periodicity investigations [51] can provide insight into seasonal influences on the temporal distribution of bacterial contamination, and seasonality has already been shown to influence results in the literature. The data are imported into MATLAB for periodicity analysis in which Fourier analysis is conducted to investigate peaks, periodic cycles and potential models. In a periodicity analysis using Fourier models, the following equation is fitted to the $LogFC_{y,m}$ versus a year-month index spanning from 1 to $n = 204$ (equal to 17 years of 12 month data), and to $LogFC_m$ versus m ($n = 12$) with the following equation:

$$f(x_t) \;=\; a_0 + \sum_{i=1}^{n} a_i \cos(i\omega t) + \sum_{i=1}^{n} b_i \sin(i\omega t) \tag{1}$$

where a_0; a_i; b_i are Fourier coefficients, and $f(x_t)$ is the predicted value of variable x_t at time t. The fundamental frequency ω is equal to $2\pi/T$ in which T is the period of the signal. The climate data are also investigated with this function in order to determine if the periods in the climate data are similar to the periods in $LogFC_{y,m}$ and $LogFC_m$.

2.3.2. Temporal Trends and Correlations between Variables

The overall correlations between $LogFC_{i,j}$ and independent variables are analyzed to determine the influence of each variable toward the level of bacterial contamination using Kendall's τ-b for non-normal distributions. While, Kendall's τ-b and the commonly used Spearman's ρ produce similar values much of the time, Kendall's test provides better explanations for non-linear correlations as compared to Spearman's test [52]. The data observed at the hot-spot stations are also tested for correlations with the independent variables. Correlations are computed between $LogFC_{i,j}$ and the following independent variables: 7-day total precipitation ($P7$), 3-day total precipitation ($P3$), 2-day total precipitation ($P2$), precipitation (P), 7-day average temperature ($T7$), 3-day average temperature ($T3$), 2-day average temperature ($T2$), mean temperature (T_{mean}), maximum temperature (T_{max}), minimum temperature (T_{min}), antecedent dry period length (t_{dry}), watershed area (WA), residential area (RA), greenspace area (GA), cloud cover (CC), and flow rate at discharge (FR).

2.3.3. Regression Modelling

Depending on the outcomes of the correlation analysis and periodicity analysis with Fourier modelling, multivariate linear regression (LR) is attempted to provide the best (in terms of coefficient of determination R^2) model possible given all the temporal data. The equations for an LR model are well known and not repeated here. In addition, Multinomial Logistic Regression modelling (MLR) is also used to explore other potential prediction models based on the limitations of LR. MLR is a method similar to LR but has the added advantage of not being constrained by several assumptions including that the independent variables must be statistically independent (although collinearity should not be great). In addition, the dependent variable may be a categorical or nominal value [53]. This was intentionally chosen because of the incredible variance in the data of FC_{ij} across the entire database of 17 years, which is also true of $LogFC_{ij}$. By using categories of $LogFC_{ij}$ determined by the order of magnitude of the $LogFC_{ij}$ value, this method can provide insight into how well individual parameters work to influence observed levels of $LogFC$, and if predictions on the likelihood of a $LogFC$ value being observed at a certain level are possible based on climatic observations. This method is considered a classification method that relates a linear prediction function to the "logit" $g(x)$ with n number of predictors x ($x = x_1, x_2, \ldots, x_n$) such that [53]:

$$g(x) = \beta_0 + \beta_1 x_1 + \beta_2 x_2 + \ldots + \beta_n x_n \tag{2}$$

and the conditional probability that the outcome Y is present is $\Pr(Y = 1|x) = \pi$ and,

$$\pi = \frac{e^{g(x)}}{1 + e^{g(x)}} \tag{3}$$

Thus, the model attempts to compute the odds, or likelihood of observing a value associated with a certain level or class. The application of the MLR model was conduct with SPSS which also provides measures of (1) model fitting information (in the form of a X^2 test that compares the model to a -2 Log Likelihood model) such that the model is significant if the p-value is <0.05; (2) goodness of fit (also at 0.05 level) relative to a baseline null model; (3) the likelihood ratio tests in which the significance of each predictor is rated as having a significant influence on the classification if the p-value is <0.05; (4) model coefficients; (5) performance in each class or level between the observed and predicted values (as a percent) as well as the overall percent performance; and (6) three Pseudo R^2 values that provide a measure of model performance [54].

In such a model, the choice of levels or classes is important. If too few classes are chosen that are too gross with little discretization, the model will perform perfectly as there are few "choices" to make, even though the model fit may not be significant. If there are too many classes, the model performance is poor. The application of this model will be on the order of magnitude of the LogFC values instead of the actual values because the actual values would create too many classes. MLR models will be developed for all of $LogFC_{i,j}$, $LogFC_{y,m}$ and for values of $LogFC_{i,j}$ at hotpsots but the "values" will now be one of 11 possible classes created as the order of the $LogFC_{ij}$ value: 2, 2.7 (representing 500 CFU/100 mL), 3, 3.7, 4, 4.7, 5, 5.7, 6, 6.7, 7, etc. The $FC_{i,j}$ values were rounded up to the nearest order of magnitude (thus, values greater than 500 CFU/100 mL, for example, would be rounded up to 1000 CFU/100 mL), then log transformed before placing them into a class. For $LogFC_{i,j}$, no intermediate classes beyond the order (i.e., no value of 2.7 for example) are chosen so there are only six classes.

It should be noted that this method has no equivalent to R^2 such as that which can be computed for a linear regression model; thus, researchers have devised what are known as Pseudo-R^2 values, with SPSS reporting three different types. Since all three provide similar values, the first reported, which is the Cox and Snell value, is given in this study and is computed as:

$$\text{Pseudo} - R^2 = 1 - \left(\frac{L_0}{L_M}\right)^{2/N} \tag{4}$$

where N are the number of observations in the dataset, L_0 is likelihood of a model with no predictors and L_M is the likelihood of the model being assessed. Great care must be taken when using this value to make interpretations. Firstly, a perfect prediction does not produce a value of 1 as in R^2, but instead for Cox and Snell's Pseudo-R^2, a perfect model performance would give a value less than 1. The literature has reported that a value close to 0.2 is considered a "good model" [55]. Since great care must be taken in the interpretation of this performance metric, MLR model performances are *compared* as opposed to examining the Pseudo-R^2 value in isolation. This is sufficient for understanding the *relative* influence of climate variables on predictions in an effort to gain knowledge where significant correlation analysis is lacking.

3. Results and Discussion

3.1. Periodicity and Fourier Models

Figure 5a,b show the time series of $LogFC_{y,m}$ and the average precipitation and average temperature in the corresponding month, along with the best fitted series for each variable. Figure 6 shows three panels with the average monthly data ($LogFC_m$ is shown in Figure 6c) over all 17 years. Fourier series models describing the fitted curves are also shown. Table 1 shows the Fourier coefficients and the RMSE and R^2 values for the best fit models. Higher order harmonics were attempted until R^2 values ceased to increase. For $P_{y,m}$ and $LogFC_{y,m}$, a simple, one harmonic best fits the data although the RMSE is high. For temperature, $T_{y,m}$, an 8 harmonic Fourier model fit the temperature function for this area over the 17 year period extremely well. The lower harmonics did not produce nearly as good an R^2 value.

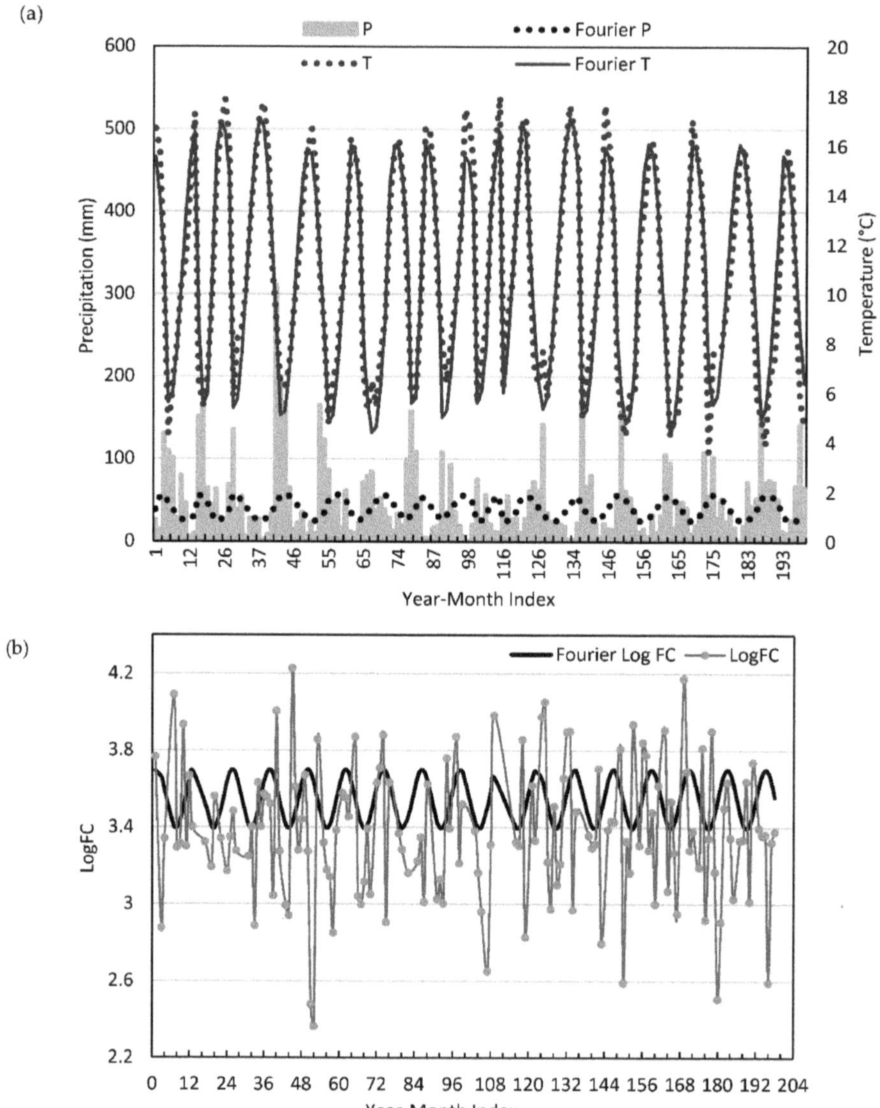

Figure 5. (**a**) Monthly climate variables and best fit Fourier series; and (**b**) $LogFC_{y,m}$ data with fitted Fourier series model.

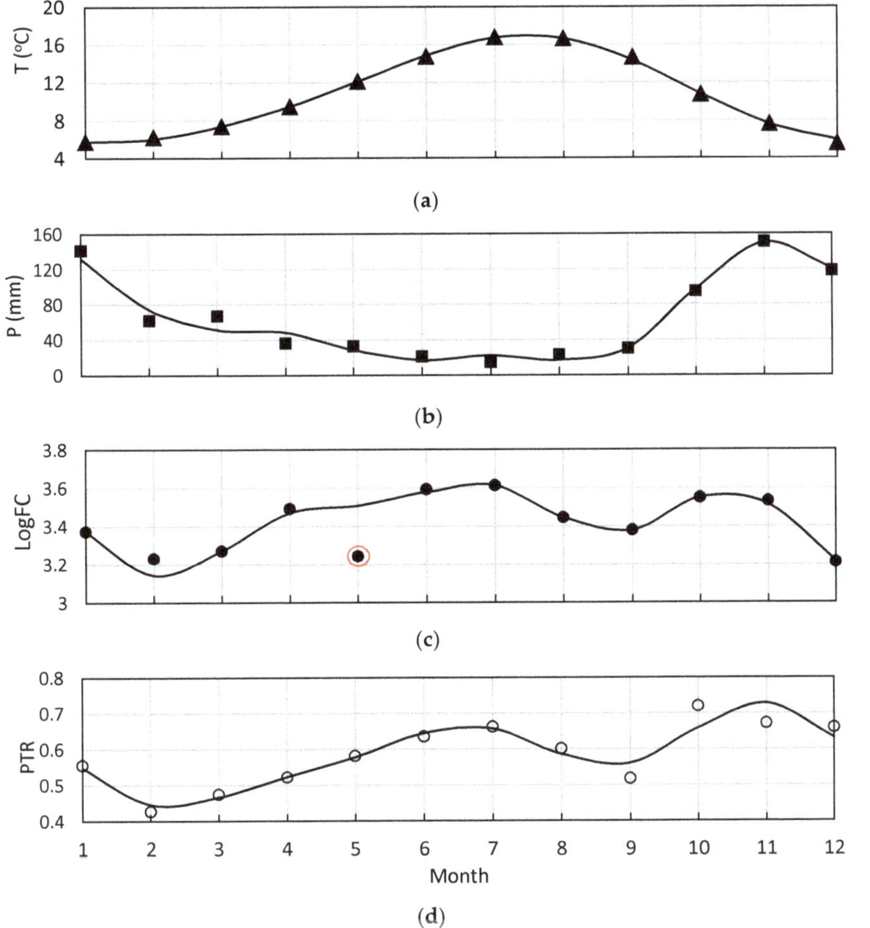

Figure 6. (**a**) Monthly temperature averaged over the entire 17 year period with fitted Fourier model shown as the solid line; (**b**) monthly averaged precipitation with Fourier model; (**c**) LogFC_m with fitted Fourier model and (**d**) PTR versus month with fitted Fourier model.

The non-uniform sampling schedule can lead to an inconsistent number of samples collected in each month, which can lead to a skewed data set in favour of times in the year when samples are taken [51]. As in [51], the positive test ratio (*PTR*) is considered in order to screen out the influence of the number of tests taken in any given month or season. *PTR* is the number of cases above the regulation limit 200 CFU/100 ml divided by the number of total cases in each month of the 17 years. Figure 6 shows the P_m, T_m, *PTR* and LogFC_m data with fitted Fourier models. Information on the fitted models are given in Table 1.

Table 1. Fourier Coefficients.

Model	Variable and Fourier Model	Adj. R^2	RMSE
1	Monthly temperature $(t = y, m)$: $= 10.72 + 0.03\cos(0.07t) + 0.41\sin(0.07t) - 0.19\cos(0.14t) - 0.16\sin(0.14t) - 0.12\cos(0.21t) + 0.05\sin(0.21t) - 0.21\cos(0.28t) + 0.12\sin(0.28t) + 0.10\cos(0.35t) + 0.08\sin(0.35t) - 0.14\cos(0.42t) - 0.07\sin(0.42t) - 0.15\cos(0.49t) - 0.21\sin(0.49t) + 4.261\cos(0.56t) + 3.648\sin(0.56t)$	0.93	1.13 °C
2	Monthly precipitation $(t = y, m)$: $= 40.6 - 9.273\cos(0.48t) + 13.32\sin(0.48t)$	0.92	15.17 mm
3	$\text{Log}FC_{y,m} = 3.547 + 0.11\cos(0.52t) + 0.11\sin(0.52t)$	0.92	0.13
4	Mean monthly temperature $(t = m) = 10.96 - 3.41\cos(0.56t) - 4.371\sin(0.56t) - 0.52\cos(1.12t) + 0.26\sin(1.12t)$	0.99	0.16 °C
5	Mean monthly precipitation $(t = m) = 59.06 + 48.6\cos(0.59t) + 3.97\sin(0.59t) + 24.07\cos(1.18t) + 7.43\sin(1.18t) + 13.69\cos(1.77t) + 5.3\sin(1.77t)$	0.92	13.95 mm
6	$\text{Log } FC_m = 3.44 - 0.6\cos(0.6t) - 0.12\sin(0.6t) + 0.11\cos(1.2t) + 0.01\sin(1.2t) + 0.09\cos(1.8t) + 0.02\sin(1.8t)$	0.69	0.08
7	$\text{PTR} = 0.59 + 0.01\cos(0.55t) + -0.07\sin(0.55t) + 0.08\cos(1.1t) - 0.02\sin(1.1t) + 0.02\cos(1.65t) - 0.03\sin(1.65t)$	0.67	0.05

In interpreting these values, what's important is the fundamental frequency ω in the dominant portion of the series. Cosine and sine terms are dominant if their coefficients are higher than other cosine and sine terms. For example, in Table 1, models 2 and 3 for precipitation and $\text{Log}FC_{y,m}$, respectively, have very similar fundamental frequencies (0.48, 0.52, respectively), which translate into both signals having a roughly 12 month periodicity in the signal. The fitted model shown in Figures 5a and 5b, respectively show the simple wave with a high R^2 value suggesting that while the peaks and troughs could not be captured in either model, both signals have this frequency. With regard to model 1, which is the 8 term Fourier model of temperature, the curves as seen in Figure 5a is nearly perfectly captured but model 1 suggests that the fundamental frequency is 0.07. However, the dominant terms in this expression are the last sine and cosine terms as they have the largest coefficients. The frequency of those terms is 0.56, again illustrating the annual periodic signal in the temperature. Since all fundamental frequencies are essentially equal, the analysis does not preclude the suggestion that $\text{Log}FC_{y,m}$ signal characteristics arise from precipitation and temperature signals.

Similar profiles are seen in Figure 6a,b for mean monthly temperature and precipitation over the entire 17 year period. Models 4 and 5 show fundamental frequencies of roughly 0.6, corresponding to a roughly annual periodicity that is clearly shown in the two panels. Models 6 and 7, for $\text{Log}FC_m$, and PTR, respectively, also show periodic signals of roughly 12 months; however, the signal peaks do not mimic either of those made by precipitation or temperature. Figure 6c shows two peaks—one in July and one in October. While both the PTR and $\text{Log}FC_m$ curves show a nearly identical curve in the fitted Fourier models, there is a glaring discrepancy in the month of May (shown circled in red in Figure 6c). Recall that the sampling program is focused on two seasons: January to April and June to September. May is not usually sampled. The PTR curve seems to suggest that the data point shown in May is an outlier requiring further consideration in the data. The models also suggest that more than just precipitation or just temperature are responsible for the oscillations seen in the curves.

The climate in southern Vancouver Island can be summarized as having less precipitation with moderate temperatures in the summer and more precipitation with cooler temperatures in the winter.

Both the minimum and maximum temperature over most of the year are not sufficient to kill the fecal coliforms since the fecal coliform could grow at temperatures between 4 °C to 35 °C and grow quickly around 20 °C [56]. The general inspection indicates temperature is one of the most important factors that cause high fecal coliform levels in the summer. During the winter time, the temperature is relatively low but the precipitation and cloud cover ratio is high and these provide moisture for fecal coliform growth and less possibility of mortality by sunlight.

The results also suggest a different seasonal pattern in fecal coliform levels from a regular monsoon climate pattern in which the level of fecal coliforms is high in summer and low in winter. The high peaks of fecal coliform concentration appear twice during the year; once is in summer around July and once in winter around October. The lowest level of fecal coliform is found in February as well as for the positive test ratio. It is apparent that the mild Mediterranean climate in Victoria could result in a relatively unusual distribution of bacterial contamination compared to other studies conducted in the world. The positive test ratio of high fecal coliform concentration is even found to be highest in October instead of July. There is also a high level of fecal coliform appearing in the samples collected in October. This finding indicates the monitoring schedule should expand outside its two season monitoring approach.

3.2. Correlation Analysis with Temporally and Spatially Related Variables in LogFC$_{i,j}$

Kendall's τ-b test results are listed in Table 2 and only those tests that returned a p-value < 0.05 are shown. In general, temperature variables and antecedent dry period have a positive correlation with the LogFC$_{i,j}$. The result shows a relatively high temperature would help fecal coliform growth in the summer and accumulate in the drainage area while no precipitation occurred. Conversely, precipitation variables have a negative correlation with the LogFC$_{i,j}$, which is unexpected as moisture helps bacteria grow and much of the literature showed an opposite correlation. The reason could be that heavy precipitation continuously washes off the bacteria and dilutes the density of fecal coliform during the wet season. The flow rate at discharge is largely negatively correlated with the fecal coliform levels, which, like precipitation suggests that higher flow rates of stormwater are diluting contaminants to the discharge. The data collection methods by the CRD intentionally avoided first flush timing, which would mean that only the late stages of the event are sampled.

Table 2. Non-parametric correlation Kendall τ-b of LogFC$_{i,j}$ for all stations and at each hot spot. The symbol '*' indicates the numeric value of τ-b has a p-value less than 0.05 and '**' indicates a p-value less than 0.01. Symbols '+' and '−' indicate the correlation was positive or negative, respectively, but is not significant.

Variable	All Stations	#805	#641	#623	#320	#245
T_{mean}	0.04 **	0.24 *	+	+	0.54 **	+
T_{min}	0.04 **	0.24 *	+	+	0.54 **	+
T_{max}	0.05 **	+	+	+	0.51 **	0.15 *
T2	0.06 **	+	+	+	0.51 **	0.16 *
T3	0.06 **	+	+	+	0.51 **	0.22 **
T7	0.07 **	+	0.10 *	+	0.51 **	0.17 *
P	−0.01 **	+	0	−	−0.19 *	−0.20 *
P2	−0.05 **	−	−	−	−0.21 *	−0.18 *
P3	−0.06 **	−0.23 *	-	-	−0.25 **	−0.25 **
P7	−0.06 **	−0.33 **	-	-	−0.38 **	−0.16 *
t_{dry}	0.06 *	+	+	-	0.26 **	0.22 *
FR	−0.05 **	−0.26 *	0.10 *	-	−0.23 **	−0.26 **
WA	+					
RA	0.11 **					
GA	+					

With regard to the land use, the residential area has a positive correlation with $LogFC_{i,j}$ which confirms the previous hypothesis that the bacteria contamination is mainly caused by human activities. The literature suggests that the watersheds with a high percentage of urban and agricultural land use mostly have high contamination levels [13]. The watershed area and the greenspace area both showed a positive but insignificant correlation suggesting that greenspace and the watershed as a whole are also contributing to the amount of fecal coliform observed. The GIS analysis shows the heavily contaminated areas are in Victoria downtown, Esquimalt and the southeastern shore of Oak Bay are densely populated and contain high percentages of impervious land use. It is reasonable to suggest that in this region, high levels of bacterial contamination are likely to be observed in populated residential areas with high percentages of impervious land use.

Table 2 also shows the correlations for selected hotspots. Station #320 showed a very strong correlation with all variables; both temperature and precipitation being significant variables with many *p*-values around 0.000. Thus, at least for this station, $LogFC$ is greatly affected by meteorological changes. This seems to be also true for station #805. The correlations shown conform to the correlation analysis of all samples from all stations: the temperature and antecedent dry period are positively correlated to $LogFC$ while the precipitation has a negative correlation. However, station #641 showed only one significant positive relationship with temperature and the opposite relationship with flowrate. While station #623 showed correlations that were in keeping with the general trend for all stations, none of them were significant. Since this sampling station is located in the nearshore of Victoria's downtown area, the distribution of fecal coliform could be dominated by spatial variables (land use) rather than temporal variability. Station #320 shows that all temperature variables are positively correlated with $LogFC$ at this sampling station and in fact, all of the relationships are significant and strong. All the temperature variables are correlated so it is not surprising that all the temperature correlations are negative, near 0.5 and all very significant (*p*-value near zero). However, there is next to no difference amongst the correlations and thus, no one variable distinguishes itself in its influence on $LogFC$ than any other temperature variable. The precipitation shows a negative correlation with $LogFC$ while the antecedent dry period has a positive correlation. The 7-day total precipitation provides the strongest negative correlation suggesting that long-term precipitation could greatly reduce the level of fecal coliforms in the area draining to this station. Station #245 shows similar trends but now the 3-day precipitation variable $P3$ has the strongest correlation for all the precipitation variables studied (i.e., $P, P2, P3, P7$). This variability in correlation strength across temperature or precipitation variables, and the lack of significance in the group of hotspot stations selected, suggests incredible variation by location in the database and that analysis on climatic/land use influences on FC or potentially prediction of FC levels should be done on a station by station basis.

The literature on fecal bacteria growth and die-off rates in the environment [56,57] suggest that the summer temperature in the study area (under 20 °C) is considered to be ideal for fecal bacteria growth. Since the maximum temperature is rarely over 30 °C, the die-off rate of fecal coliform is also low in the study area. McCarthy et al. [27] suggested that temperature was an important variable such that E. coli could be persisting or growing in the catchment during periods of warmer temperature, and this may be applicable in this study region.

The correlation analysis shows that precipitation is negatively correlated with fecal coliform concentration, but the relationship is weaker than that for temperature. The negative correlation contradicts some of the literature [58] but seasonal dependence is important [26,30]. The 7-day total precipitation has the strongest negative correlation coefficient among the precipitation variables in two of the stations, which may suggest that long-term rainfall could further decrease the fecal coliform level of the area. The southern Vancouver Island region usually has light and long-term precipitation during the wet season. With regard to the positive correlation of $LogFC$ and antecedent dry period, this could be due to long inter-event periods allowing significant accumulation of fecal coliform in the drainage area. When storms occur, because they are not sampled at the first flush, there will be dilution of the fecal coliform concentration. However, one needs to note that the chances of obtaining a sample with over the limit bacteria concentrations is even higher during the winter (as noted by the curve of PTR in Figure 6d) even if the average value of fecal coliform is lower than in the summer. This indicates that the minimum temperature (approximately 3 °C) in the study area is not sufficient to inactivate fecal bacteria, and the bacterial contaminants could be transported by stormwater to the discharge in the cooler, wetter season.

Table 3 shows the correlations when comparing $LogFC_{y,m}$ and the available temporal datasets including cloud cover CC. This table shows that all the temperature variables and the cloud cover variable have a significant influence on $LogFC_{y,m}$ with T_{min} having the strongest correlation of the variables, albeit, the difference is very small. This would suggest that temperature could be a useful predictor of $LogFC$ for general planning purposes on an annual basis, or in climate change studies.

Table 3. Non-parametric correlation Kendall τ-b of $LogFC_{y,m}$ and p-values.

Statistics	T_{min}	T	T_{max}	P	CC
τ-b correlation	0.122	0.121	0.119	−0.092	−0.117
p-value	0.024	0.025	0.027	0.088	0.037

3.3. Climate Related Modelling

3.3.1. Multiple Linear Regression

Through the previous correlation analysis process, the fecal coliform value has shown a significant relationship with meteorological variability, although the correlations are weak. The periodicity modelling showed that there are issues raised by PTR versus the month averaged over all 17 years. Figure 6c is the data with a Fourier model superimposed on the $LogFC_m$ data showing a circled May data point. This data point in Figure 6d corresponds to the PTR data which shows no uncharacteristic dip but follows the 3 term Fourier model much like the $LogFC_m$ Fourier model. Since the data sampling period strongly skews data in favour of the months of January to April (henceforth, referred to as Season 1) and June to September (Season 2), the month of May does not have a lot of data points but interestingly, the data point in Figure 6c for May seems uncharacteristically out of place. The months of October to December have some data but are also sparse over the 17 year period. Therefore, any further climate analysis examines the data for $LogFC_{i,j}$ with the months of May, and October to December removed. The database of the full 17 year period now spans 2181 data points down from 2365 (after the sewage suspected samples were removed as indicated earlier).

The goal of this section is to develop a general function for predicting the $LogFC$ with selected meteorological variables. The significant but low correlations suggest that attempting to linearly regress all of $LogFC_{i,j}$ with one or more climate variables may likely show a significant relationship but the R^2 would likely be poor. Figures 7a, 7b and 7c show the data as a function of mean temperature for all $LogFC_{i,j}$, $LogFC_{y,m}$ and $LogFC_m$, respectively. These scatterplots indicate the variance in the data that a linear regression model must attempt to explain. Thus, LR was only attempted on $LogFC_m$.

The LR modelling was conducted through SPSS and uses Akaike information criterion to avoid potential overfitting and only chose the most significant variables. The variables tested were all those listed in Table 2 (with the exception of the watershed information, and the flowrate when it was not possible) and cloud cover (CC) was also included. The variables providing the best fit are T_{min} and CC with an R^2 of 0.55 and an adjusted R^2 of the model shown in Equation (5) as 0.45. The RMSE = 0.109 and the significance of the T_{min} and CC coefficients are 0.02, and 0.07, respectively.

$$LogFC = 2.05 + 0.08T_{min} + 1.41CC. \tag{5}$$

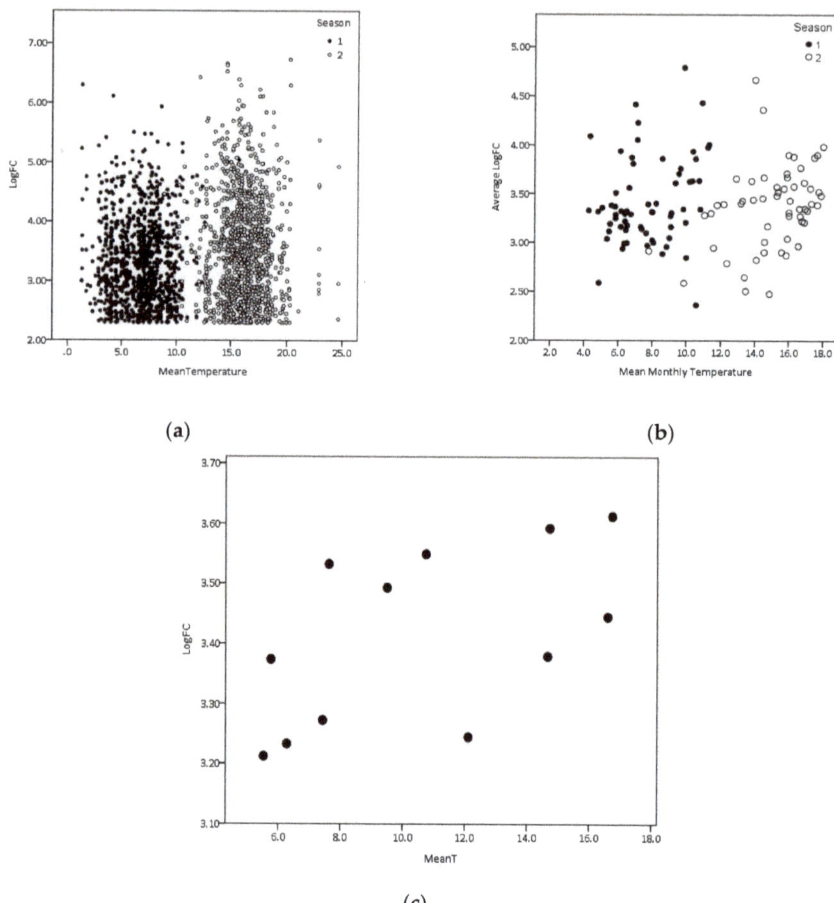

Figure 7. Scatterplots of (**a**) $LogFC_{i,j}$ versus temperature of the day of sampling at each location; (**b**) $LogFC_{y,m}$ versus mean monthly temperature; and (**c**) $LogFC_m$ versus mean monthly temperature over the entire period in each month.

Notice that in Equation (5), the coefficient multiplied by CC is positive not negative. When no weather or obstructions to visibility occur, sky conditions are provided reflecting the observation of total cloud amount. The value of CC based on the amount (in tenths) of cloud covering the dome of the sky with 0 being clear and 1.0 being completely clouded over sky. While higher cloud clover might suggest greater opportunity for precipitation, which is negatively correlated with FC in this region, in the absence of precipitation, cloud cover would prevent solar radiation from promoting bacteria die-off. This concurs with [29] that stated that higher solar radiation lead to lower bacteria levels.

While the R^2 is only fair, the model coefficient for T_{min} is significant and nearly significant for CC. This is not surprising given the results of the correlation analysis. In addition, some of the principles behind the non-parametric correlation analysis are similar to some of the assumptions underlying multiple linear regression. The Fourier analysis suggested a seasonal influence on the data and the differences in correlations in Table 2 suggest that location in space is important. Thus, averaging across space and over the year, as is done for $LogFC_m$ in this analysis, would produce large errors in any prediction model. Figure 8 shows plots of the residuals in the predict $LogFC_{y,m}$ values. The residuals for T_{min} are scattered over the temperature range experienced over the year and there is no trend that warmer or cooler temperatures lead to better predictions. This is also true of cloud cover in that the residuals are scattered around 0 for the full range of *CC*.

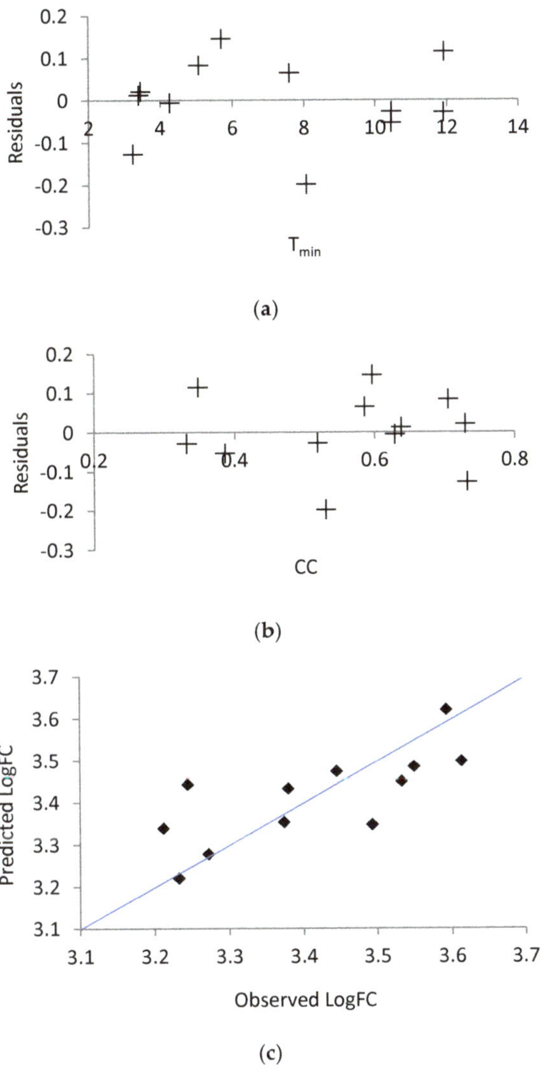

(a)

(b)

(c)

Figure 8. Residuals as a function of (**a**) T_{min}, and (**b**) *CC*, and (**c**) trend of predicted versus observed $LogFC_m$ with one to one line shown in blue.

3.3.2. Multinomial Logistic Regression in $LogFC_{y,m}$

Tables 4–6 show the results of the MLR on $LogFC_{i,j}$, $LogFC_{i,j}$ at the hotspots and $LogFC_m$, respectively. Each variable is tested for its significance in explaining the likelihood of LogFC belonging to a certain order of magnitude (class), and then the percentage of model performance for each class is computed. The shaded cells indicate that the predictor had a significant influence on the overall model performance. The model performances are calculated as a weighted performance in each class. If no value of performance is shown for a class that means that there were no observations for that class in the data set. The number of data points in the set are indicated by N and whether or not the model fit is significant or not is shown with shading (no shading indicates it is not significant).

Table 4. MLR modelling results for $LogFC_{i,j}$ for all stations. A shaded box indicates a significant relationship with no shading indicating the p-value was greater than 0.05.

	Period	All	Season 1	Season 2	Month 1	2	3	4	6	7	8	9
Qual-ity	N	2181	1038	953	307	424	216	91	157	419	357	210
	Model Fit Pseudo-R^2	0.06	0.10	0.06	0.17	0.13	0.31	0.49	0.34	0.19	0.21	0.32
Predictor Significance	Intercept											
	Flow Rate											
	T											
	T_{min}											
	T_{max}											
	T2											
	T3											
	T7											
	P											
	P2											
	P3											
	P7											
	t_{dry}											
Class of LogFC	Class					% Performance						
	2	41	57	42	59	71	68	58	53	45	55	65
	3	64	52	70	56	47	51	67	72	69	62	35
	4	0	0	4	10	0	16	13	128	16	17	47
	5	0	0	0	0	0	0	100	0	0	5	33
	6	0	100	0		100	100		100	0	100	100
	Overall	41	46	42	52	50	51	56	48	44	47	50

In examining Table 4 closely, for all the 2181 records of $LogFC_{i,j}$, flowrate and seven day average temperature and flowrate were found to significantly influence the likelihood of observing an order of LogFC but the overall model performance was only 41%. When examining the difference introduced by seasonality, Season 1 data provided a better model performance than Season 2 and was able to perfectly predict the values of order of magnitude 6 and higher. However, it could only predict values in order 2 and 3, which were the bulk of the database with only roughly 50% performance. Here again flowrate was significant but now so was temperature T and T2 and P and P2. Season 2 showed a very good performance for order 3—better than in Season 1 but now only T3 and T7 were significantly influencing results indicating that only temperature matters in the warmer season. This is not surprising as this is also the drier season. However, flowrate was still significantly influencing predictions in Season 2 as well. This however changed when focusing on the monthly performance.

The analysis by month showed that flowrate was significant in August and September—two of the four months of the warm season but the peak in FC levels occurs in July which does not have flowrate significantly influencing the model performance. The month of February showed absolutely no significant influence of any variable. The peak month of July showed that maximum observed temperature and precipitation on the day or within two days of sampling were significantly influencing the model results. July also had the lowest performance as compared to the other months in Season 2, which were all able to predict occurrences of order 6 and higher in $LogFC_{i,j}$ at 100%. Interestingly, September had the best spread of performances in each class for Season 2 and here flowrate and seven day accumulation of precipitation were key to that performance. The best performance overall was for the month of April in which, flowrate, T3, T7, and t_{dry} were key. The Pseudo-R^2 values were only above 0.2 in the months of March, April, June, August and September, again suggesting that monthly time scales for examining climate influences is warranted.

Table 5. MLR modelling results for Log$FC_{i,j}$ for the hotspot stations. A shaded box indicates a significant relationship with no shading indicating the p-value was greater than 0.05. The letter "B" indicates the relationship is not significant but the p-value was very close to 0.05.

Period	Season			Month							
	All	**1**	**2**	**1**	**2**	**3**	**4**	**6**	**7**	**8**	**9**
N	210	99	111	33	28	30	8	22	33	29	27
Model Fit		B									B	
Pseudo R^2	0.52	0.60	0.66	0.96	0.94	0.93	0.86	0.97	0.94	0.92	0.54
Intercept											
Flow Rate											B	
T											
T_{min}											
T_{max}											
T2					B						
T3											
T7											
P											
P2											
P3											
P7			B								
t_{dry}											
Class					Performance in (%)							
2	100	100	100	100		0						0
2.70	6	20	17	100	100	67		100	0	0.5	100
3	38	50	39	100	100	71	100	100	100	100	100
3.70	54	57	66	100	83	25	100	100	83	92	71
4	21	44	33	100	100	83	100	100	88	50	25
4.70	50	47	48	100	100	86		100	80	80	100
5	38		36					100	100	100	67
5.70	100		50					100			100
6	75	100	100			100			100		
6.70	100		100					100			
Overall	41	50	48	100	96	70	100	100	84	86	74

Table 6. MLR modelling results for Log$FC_{y,m}$. A shaded box indicates a significant relationship and no shading indicates the p-value was greater than 0.05.

	All	**Season 1**	**Season 2**
N	116	59	57
Model Fit			
Pseudo R^2	0.40	0.29	0.52
Intercept			
T_{min}			
T			
T_{max}			
P			
CC			
Class	% Performance (N in class)		
2.7	0	100 (1)	0 (2)
3	77	95 (41)	52 (21)
3.7	60	27 (11)	77 (26)
4	8	0 (6)	33 (6)
4.7	100		100 (2)
Overall Performance	63	73	61

Table 5 shows the results for the hotspot stations. Recall from Table 2, only two to three of the five hotspot stations had climatic variables and land use variables with significant effects on $LogFC_{i,j}$. In Table 5, the overall performance for all the data, Seasons 1 and 2, are similar to the entire dataset but now, for the whole set, flowrate and precipitation are influencing results as opposed to flowrate and temperature. The only variable influencing performance in Season 1 is seven day precipitation and only marginally in Season 2, in which the model fit was not significant. A more granular analysis by month showed that the two months with high levels of FC, June and July were nicely modelled with flowrate and $T3$ providing a perfect model performance at all classes in the month of June. However, the month of July, had a poorer performance but flowrate was not a factor but other temperature variables, precipitation variables and dry days were statistically relevant. The Cox and Snell Pseudo-R^2 values are all greater than 0.2 suggesting good model performances in all variables used in the model.

Table 6 shows the results for the 116 data points without May, October–December values seen in Figure 5b. Again, the data were also examined by season but data by month were sparse, which would lead to falsely inflated values of model performance and thus, were not examined.

The model performance overall is the best seen for all data sets which is not surprising. As well, while Season 1 had the best performance, the fit was not significant and no single predictor had a significant influence. The overall model performance in Season 2 was 61% with good performance in the middle and higher order classes. Here all the variables were significant with the exception of cloud cover, which proved useful in the multiple linear regression modelling for the data set of Figure 6c (monthly averages over the entire database $LogFC_m$). In the logit model of Equation (3), the model coefficients β are specific to each class.

To illustrate the model implications on the relationship between increases or decreases in fundamental variables, two classes are created for this dataset: $LogFC < 3.7$ assigned to order 3 and those ≥ 3.7 are assigned to order 4. The results are shown in Table 7. In this analysis, category 4 was the reference category, and P and T were the selected variables. Unlike the LR modelling of Section 3.3.1, CC had no impact on the results. As T_{min} and T_{max} are certainly related to T, and to understand the general influence of T and P as two distinct climatic variables on $LogFC_{y,m}$, these two variables were selected for the analysis in this illustration.

Table 7. MLR modelling results for $LogFC_{y,m}$ with only classes 3 and 4 and variables, T and P.

N	116
Model Fit Significant?	Y
Goodness of Fit Significant?	Y
Pseudo R^2	0.12
Intercept β_0	−0.614
β of T	0.078
β of P	−0.013
3	68 (65)
4	67 (51)
Overall Performance	67%

In terms of the logistic model, these coefficients suggest that,

$$\text{logit}(\pi) = -0.614 + 0.078T - 0.013P \tag{6}$$

In Equation (6) each coefficient is the expected change in the log odds of observing a $LogFC$ of magnitude 3 for a unit increase in a predictor variable when the other predictor variables are fixed. The value of e^β for a specific predictor is the change in odds (in the multiplicative scale) for a unit increase in that predictor. The intercept, in this case, is the log odds of observing a value of order of magnitude 3 if temperature and precipitation are zero because that was chosen as the reference category in this analysis. This fitted model says that if precipitation is held fixed, the coefficient for temperature suggests that for a unit increase in temperature (1 °C) would result in an increase in the odds of observing a Log*FC* with order of magnitude of 4 by 8% (since $e^{0.078} = 1.08$). Conversely, if temperature is held fixed, a unit increase in precipitation (1 mm) would result in a decrease in the odds of observing an order of magnitude 4 by 2%.

This analysis reaffirms the suggestion that in this study region, temperature increases are likely to increase Log*FC* levels and precipitation likely lowers Log*FC* levels. The results in Tables 6 and 7 all involve a spatial averaging over the study area, which has already suggested possible errors in modelling. Vermuelen and Hofstra [31] noted that temperature was a significant variable in four of six locations studied yet leaving it out of their overall model did not make a "great difference" for a model with an adjusted R^2 of 0.49. The implication of spatial influences was further iterated by Mallin et al. [35] who found that multiple regression models failed to explain any more of the variability in bacteria than simply using percent impervious surface cover alone. Seasonality implications were clearly visible in the results. In Selvakumar and Borst [59], the authors found seasonal variations in the data in which early season storms generally had higher levels of bacteria than late season storms, both within and between watersheds. But when spatial implications were ignored and all watersheds were lumped together, bacteria levels decreased with increasing cumulative annual rainfall. This suggests that if detailed modelling or planning is going to be conducted, it should not only be conducted with attention to location, but attention to season.

4. Conclusions

This study provides insight into the temporal and spatial distribution of bacterial contamination in nearshore areas of southern Vancouver Island and explains the potential relationships in fecal coliform levels in stormwater with land-use and climate variables. Non-parametric correlation analysis, Fourier analysis, multiple linear regression and multivariate logistic regression were used to determine the significance and relative influence of climatic variables on Log*FC* levels measured between 1995 and 2011 at three temporal scales: at the time of sampling, the monthly average of Log*FC* across all stations in the region, and the average across space and time for each month.

The Fourier model demonstrates an annual periodicity of fecal coliform with two peaks in July and October. Despite the extreme values observed in the 17 years' historical data, the data follow the general trend of this periodicity as do the precipitation and temperature. No matter the temporal scale used, the Fourier analysis always produced a periodicity of 12 months.

The correlation analysis showed that fecal coliform concentrations for all stations throughout the period have a significant correlation with many climatic variables including positive correlations with temperature, antecedent dry period and land use in the watershed, and negative correlations with precipitation and flowrate. Drainage area showed no significant correlations with bacterial contamination. But drainage area, amount of residential area in the drainage area as well as green space were all positively correlated with FC levels but only the amount of residential, or urban area, was significantly correlated. This suggests that the bacteria loading mostly originates from residential activity. The incredible variability in whether relationships were significant or not suggest that modelling or planning should be done on a location basis and not necessarily averaged across the region. Seasonal effects were also apparent and planning and prediction should be conducted differently in the two seasons. In particular, the drier, hotter season had higher levels of *FC* and therefore, should be the focus of future planning.

Linear regression developed a prediction model for Log*FC* in each month with a low adjusted R^2 of under 0.5, but with significant or marginally significant variables of minimum monthly temperature and cloud cover. Multivariate logistic regression was used with classes representing the order of magnitude of Log*FC*. This type of regression showed very good model performance, which improved when considering the wet season separately from the dry season, and when looking at a individual months.

The authors recommend that when examining the influences of climate change on fecal coliforms, and possibly, on other types of bacterial contaminants in this region, that variables be distinguished by season, and possibly by month, as well as by location. In addition, the authors recommend that the CRD monitor FC more intensively at every month of the year, and in the fall season that includes October. The analysis should be updated whenever new data are made available. Since the land use was only represented by only two different types of land use (residential versus greenspace), the study should be expanded to other areas of the region that include forested and agricultural lands. The MLR model could be easily applied on the updated and expanded datasets.

Author Contributions: Conceptualization, C.V.; Methodology, C.V. and K.X.; Formal Analysis, K.X. and Z.X.; Investigation, K.X.; Data Curation, K.X. and Z.X.; Writing-Original Draft Preparation, K.X.; Writing-Review & Editing, C.V. and J.H.; Supervision, C.V. and J.H.

Funding: This research received no external funding.

Acknowledgments: The Authors would like to thank Natalie Bandringa and Dale Green of the Capital Regional District for the support of this work; Barri Rudolph of the CRD for supplying data; and Joachim Carolsfeld of World Fish Organization for his support.

Conflicts of Interest: The authors declare no conflict of interest.

References

1. Bowen, R.; Depledge, M. Rapid Assessment of Marine Pollution (RAMP). *Mar. Pollut. Bull.* **2006**, *53*, 631–639. [CrossRef]
2. Stormwater, Harbours and Watersheds Program Environmental Sustainability. In *Core Area Stormwater Quality 2012 Annual Report*; Capital Regional District: Victoria, BC, Canada, 2013; Available online: http://www.crd.bc.ca/about/what-we-do/stormwater-wastewater-septic/monitoring-stormwater (accessed on 1 January 2014).
3. Curriero, F.; Patz, J.; Rose, J.; Lele, S. The Association Between Extreme Precipitation and Waterborne Disease Outbreaks in the United States, 1948–1994. *Am. J. Public Health* **2001**, *91*, 1194–1199. [CrossRef]
4. Gaffield, S.; Goo, R.; Richards, L.; Jackson, R. Public Health Effects of Inadequately Managed Stormwater Runoff. *Am. J. Public Health* **2003**, *93*, 1527–1533. [CrossRef] [PubMed]
5. Ahmed, W.; Goonetilleke, A.; Gardner, T. Human and bovine adenoviruses for the detection of source-specific fecal pollution in coastal waters in Australia. *Water Res.* **2010**, *44*, 4662–4673. [CrossRef] [PubMed]
6. Frenzel, S.; Couvillion, C. Fecal-indicator bacteria in streams along a gradient of residential development. *J. Am. Water Resour. Assoc.* **2002**, *38*, 265–273. [CrossRef]
7. Campos, C.; Cachola, R. Faecal Coliforms in Bivalve Harvesting Areas of the Alvor Lagoon (Southern Portugal): Influence of Seasonal Variability and Urban Development. *Environ. Monit. Assess.* **2007**, *133*, 31–41. [CrossRef] [PubMed]
8. Stocker, M.; Rodriguez-Valentin, J.; Pachepsky, Y.; Shelton, D. Spatial and temporal variation of fecal indicator organisms in two creeks in Beltsville, Maryland. *Water Qual. Res. J. Can.* **2016**, *51*, 167–179. [CrossRef]
9. Davis, K.; Anderson, M.; Yates, M. Distribution of indicator bacteria in Canyon Lake, California. *Water Res.* **2005**, *39*, 1277–1288. [CrossRef] [PubMed]
10. Sibanda, T.; Chigor, V.; Okoh, A. Seasonal and spatio-temporal distribution of faecal-indicator bacteria in Tyume River in the Eastern Cape Province, South Africa. *Environ. Monit. Assess.* **2012**, *185*, 6579–6590. [CrossRef]
11. Hunter, C.; Perkins, J.; Tranter, J.; Gunn, J. Agricultural land-use effects on the indicator bacterial quality of an upland stream in the Derbyshire peak district in the UK. *Water Res.* **1999**, *33*, 3577–3586. [CrossRef]
12. Crowther, J.; Kay, D.; Wyer, M. Faecal-indicator concentrations in waters draining lowland pastoral catchments in the UK: Relationships with land use and farming practices. *Water Res.* **2002**, *36*, 1725–1734. [CrossRef]
13. Tong, S.; Chen, W. Modeling the relationship between land use and surface water quality. *J. Environ. Manag.* **2002**, *66*, 377–393. [CrossRef] [PubMed]

14. Traister, E.; Anisfeld, S. Variability of Indicator Bacteria at Different Time Scales in the Upper Hoosic River Watershed. *Environ. Sci. Technol.* **2006**, *40*, 4990–4995. [CrossRef] [PubMed]

15. Bolstad, P.; Swank, W. Cumulative impacts of landuse on water quality in a southern Appalachian watershed. *J. Am. Water Resour. Assoc.* **1997**, *33*, 519–533. [CrossRef]

16. Chu, Y.; Tournoud, M.; Salles, C.; Got, P.; Perrin, J.; Rodier, C.; Caro, A.; Troussellier, M. Spatial and temporal dynamics of bacterial contamination in South France coastal rivers: Focus on in-stream processes during low flows and floods. *Hydrol. Process.* **2013**, *28*, 3300–3313. [CrossRef]

17. Wu, J.; Rees, P.; Dorner, S. Variability of *E. Coli* Density and Sources in an Urban Watershed. *J. Water Health* **2011**, *9*, 94–106. [CrossRef]

18. Henry, R.; Schlang, C.; Coutts, S.; Kolotelo, P.; Prosser, T.; Crosbie, N.; Grant, T.; Cottam, D.; O'Brien, P.; Deletic, A.; et al. Into the deep: Evaluation of SourceTracker for assessment of faecal contamination of coastal waters. *Water Res.* **2016**, *93*, 242–253. [CrossRef]

19. Jent, J.R.; Ryu, H.; Toledo-Hernández, C.; Santo Domingo, J.W.; Yeghiazarian, L. Determining Hot Spots of Fecal Contamination in a Tropical Watershed by Combining Land-Use Information and Meteorological Data with Source-Specific Assays. *Environ. Sci. Technol.* **2013**, *47*, 5794–5802. [CrossRef]

20. Liang, Z.; He, Z.; Zhou, X.; Powell, C.A.; Yang, Y.; He, L.M.; Stofella, P.J. Impact of Mixed Land-Use Practices on the Microbial Water Quality in a Subtropical Coastal Watershed. *Sci. Total Environ.* **2013**, *449*, 426–433. [CrossRef]

21. Tiefenthaler, L.; Stein, E.D.; Schiff, K.C. Levels and Patterns of Fecal Indicator Bacteria in Stormwater Runoff from Homogenous Land Use Sites and Urban Watersheds. *J. Water Health* **2011**, *9*, 279–290. [CrossRef]

22. Walters, S.P.; Thebo, A.L. *Water Res.* **2010**, *45*, 1752–1762. [CrossRef] [PubMed]

23. Wu, J.; Yunus, M.; Islam, M.S.; Emch, M. Influence of Climate Extremes and Land Use on Fecal Contamination of Shallow Tubewells in Bangladesh. *Environ. Sci. Technol.* **2016**, *50*, 2669–2676. [CrossRef] [PubMed]

24. Cha, Y.; Park, M.H.; Lee, S.H.; Kim, J.H.; Cho, K.H. Modeling Spatiotemporal Bacterial Variability with Meteorological and Watershed Land-Use Characteristics. *Water Res.* **2016**, *100*, 306–315. [CrossRef] [PubMed]

25. Paule-Mercado, M.A.; Ventura, J.S.; Memon, S.A.; Jahng, D.; Kang, J.H.; Lee, C.H. Monitoring and Predicting the Fecal Indicator Bacteria Concentrations from Agricultural, Mixed Land Use and Urban Stormwater Runoff. *Sci. Total Environ.* **2016**, *550*, 1171–1181. [CrossRef]

26. Cho, K.H.; Pachepsky, Y.A.; Kim, J.H.; Guber, A.K.; Shelton, D.R.; Rowland, R. Release of *Escherichia Coli* from the Bottom Sediment in a First-Order Creek: Experiment and Reach-Specific Modeling. *J. Hydrol.* **2010**, *391*, 322–332. [CrossRef]

27. McCarthy, D.T.; Hathaway, J.M.; Hunt, W.F.; Deletic, A. Intra-Event Variability of *Escherichia Coli* and Total Suspended Solids in Urban Stormwater Runoff. *Water Res.* **2012**, *46*, 6661–6670. [CrossRef]

28. Malin, M.; Cahoon, L.; Parsons, D.; Ensign, S. Effect of Nitrogen and Phosphorus Loading in Coastal Plain Blackwater Rivers. *J. Freshw. Ecol.* **2001**, *16*, 455–464. [CrossRef]

29. Cho, K.H.; Cha, S.M.; Kang, J.H.; Lee, S.W.; Park, Y.; Kim, J.W.; Kim, J.H. Meteorological Effects on the Levels of Fecal Indicator Bacteria in an Urban Stream: A Modeling Approach. *Water Res.* **2009**, *44*, 2189–2202. [CrossRef]

30. Jung, Y.; Park, Y.; Lee, K.; Kim, M.; Go, K.; Park, S.; Kwon, S.; Yang, J.; Mok, J. Spatial and seasonal variation of pollution sources in proximity of the Jaranman-Saryangdo area in Korea. *Mar. Pollut. Bull.* **2017**, *115*, 369–375. [CrossRef]

31. Vermeulen, L.C.; Hofstra, N. Influence of Climate Variables on the Concentration of *Escherichia Coli* in the Rhine, Meuse, and Drentse Aa during 1985–2010. *Reg. Environ. Chang.* **2014**, *14*, 307–319. [CrossRef]

32. Martinez-Urtaza, J.; Saco, M.; de Novoa, J.; Perez-Pineiro, P.; Peiteado, J.; Lozano-Leon, A.; Garcia-Martin, O. Influence of Environmental Factors and Human Activity on the Presence of Salmonella Serovars in a Marine Environment. *Appl. Environ. Microbiol.* **2004**, *70*, 2089–2097. [CrossRef] [PubMed]

33. Patz, J.; Vavrus, S.; Uejio, C.; McLellan, S. Climate Change and Waterborne Disease Risk in the Great Lakes Region of the U.S. *Am. J. Prev. Med.* **2008**, *35*, 451–458. [CrossRef] [PubMed]

34. Arnone, R.; Walling, J. Waterborne pathogens in urban watersheds. *J. Water Health* **2007**, *5*, 149–162. [CrossRef] [PubMed]

35. Mallin, M.A.; Williams, K.E.; Esham, C.; Lowe, R.P. Effect of Human Development on Bacteriological Water Quality in Coastal Watersheds. *Ecol. Appl.* **2000**, *10*, 1047–1056. [CrossRef]

36. Vitro, K.A.; BenDor, T.K.; Jordanova, T.V.; Miles, B. A Geospatial Analysis of Land Use and Stormwater Management on Fecal Coliform Contamination in North Carolina Streams. *Sci. Total Environ.* **2017**, *603–604*, 709–727. [CrossRef] [PubMed]

37. Delpla, I.; Rodriguez, M.J. Effects of Future Climate and Land Use Scenarios on Riverine Source Water Quality. *Sci. Total Environ.* **2014**, *493*, 1014–1024. [CrossRef]

38. St. Laurent, J.; Mazumder, A. The influence of land-use composition on fecal contamination of riverine source water in southern British Columbia. *Water Resour. Res.* **2012**, *48*. [CrossRef]

39. Galfi, H.; Österlund, H.; Marsalek, J.; Viklander, M. Indicator Bacteria and Associated Water Quality Constituents in Stormwater and Snowmelt from Four Urban Catchments. *J. Hydrol.* **2016**, *539*, 125–140. [CrossRef]

40. Bravo, H.; McLellan, S.; Klump, J.; Hamidi, S.; Talarczyk, D. Modeling the fecal coliform footprint in a Lake Michigan urban coastal area. *Environ. Model. Softw.* **2017**, *95*, 401–419. [CrossRef]

41. Kuppusamy, M.; Giridhar, V. Factor analysis of water quality characteristics including trace metal speciation in the coastal environmental system of Chennai Ennore. *Environ. Int.* **2006**, *32*, 174–179. [CrossRef]

42. Simeonov, V.; Stratis, J.; Samara, C.; Zachariadis, G.; Voutsa, D.; Anthemidis, A.; Sofoniou, M.; Kouimtzis, T. Assessment of the surface water quality in Northern Greece. *Water Res.* **2003**, *37*, 4119–4124. [CrossRef]

43. Singh, K.; Malik, A.; Mohan, D.; Sinha, S. Multivariate statistical techniques for the evaluation of spatial and temporal variations in water quality of Gomti River (India)—A case study. *Water Res.* **2004**, *38*, 3980–3992. [CrossRef] [PubMed]

44. Huang, J.; Ho, M.; Du, P. Assessment of temporal and spatial variation of coastal water quality and source identification along Macau peninsula. *Stoch. Environ. Res. Risk Assess.* **2010**, *25*, 353–361. [CrossRef]

45. Marmontel, C.V.F.; Lucas-Borja, M.E.; Rodrgues, V.A.; Zema, D.A. Effects of land use and sampling distance on water quality in tropical headwater springs (Pimenta cree, Sao Paulo State, Brazil). *Sci. Total Environ.* **2018**, *622*, 690–701. [CrossRef] [PubMed]

46. Khan, U.T.; Valeo, C. Comparing a Bayesian and fuzzy number approach to uncertainty quantification in short-term dissolved oxygen prediction. *J. Environ. Inform.* **2017**, *30*, 1–16. [CrossRef]

47. He, J.; Chu, A.; Ryan, M.C.; Valeo, C.; Zaitlyn, B. Abiotic influences on dissolved oxygen in a riverine environment. *Ecol. Eng.* **2011**, *37*, 1804–1814. [CrossRef]

48. Pike, R.G.; Redding, T.E.; Moore, R.D.; Winkler, R.D.; Bladon, K.D. *Compendium of Forest Hydrology and Geomorphology in British Columbia*; B.C. Ministry of Forests and Range, Government Publications Services: Victoria, BC, Canada, 2010; ISBN 978-0-7726-6332-0.

49. *Guidelines for Canadian Recreational Water Quality*, 3rd ed.; Catalogue No H129-15/2012E; Health Canada: Ottawa, ON, Canada, 2012; Available online: http://www.healthcanada.gc.ca/waterquality (accessed on 1 January 2014).

50. Bailey, T.C.; Gatrell, A.C. *Interactive Spatial Data Analysis*; Longman Scientific Technical: Essex, UK, 1995.

51. Valeo, C.; Checkley, S.; He, J.; Neumann, N. Rainfall and microbial contamination in Alberta well water. *J. Environ. Eng. Sci.* **2016**, *11*, 18–28. [CrossRef]

52. Hauke, J.; Kossowski, T. Comparison of Values of Pearson's and Spearman's Correlation Coefficients on the Same Sets of Data. *Quaest. Geogr.* **2011**, *30*. [CrossRef]

53. Hosmer, D.W.; Lemeshow, S.; Sturdivant, R.X. *Applied Logistic Regression*, 3rd ed.; John Wiley Sons, Inc.: Hoboken, NJ, USA, 2013; ISBN 978-0-470-58247-3.

54. Cleophas, T.; Zwinderman, A. *SPSS for Starters and 2 Levelers*; Springer International Publishing: Cham, Switzerland, 2016; ISBN 978-3-319-20600-4.

55. Tabachnick, B.G.; Fidell, L.S. *Using Multivariate Statistics*, 6th ed.; Pearson: New York, NY, USA, 2012; p. 983. ISBN 978-0205849574.

56. Guber, A.; Fry, J.; Ives, R.; Rose, J. *Escherichia coli* Survival in, and Release from, White-Taid Deer Feces. *Appl. Environ. Microbiol.* **2014**, *81*, 1168–1176. [CrossRef]

57. Selvakumar, A.; Borst, M.; Struck, S. Microorganisms Die-Off Rates in Urban Stormwater Runoff. *Proc. Water Environ. Fed.* **2007**, *5*, 214–230. [CrossRef]

58. Molina, M.; Hunter, S.; Cyterski, M.; Peed, L.A.; Kelty, C.A.; Sivagensan, M.; Mooney, T.; Prieto, L.; Shanks, O.C. Factors Affecting the Presence of Human-Associated and Fecal Indicator Real-Time Quantitative PCR Genetic Markers in Urban-Impacted Recreational Beaches. *Water Res.* **2014**, *64*, 196–208. [CrossRef] [PubMed]
59. Selvakumar, A.; Borst, M. Variation of Microorganism Concentration in Urban Stormwater Runoff with Land Use and Seasons. *J. Water Health* **2006**, *4*, 109–124. [CrossRef] [PubMed]

Article

Effects of Extensive Green Roofs on Energy Performance of School Buildings in Four North American Climates

Milad Mahmoodzadeh [1], Phalguni Mukhopadhyaya [1] and Caterina Valeo [2,*]

[1] Civil Engineering, University of Victoria, Victoria, BC V8W 2Y2, Canada; miladmahmoodzadeh@uvic.ca (M.M.); phalguni@uvic.ca (P.M.)
[2] Mechanical Engineering, University of Victoria, Victoria, BC V8W 2Y2, Canada
* Correspondence: valeo@uvic.ca; Tel.: +1-250-721-8623

Received: 18 September 2019; Accepted: 11 December 2019; Published: 18 December 2019

Abstract: A comprehensive parametric analysis was conducted to evaluate the influence of the green roof design parameters on the thermal or energy performance of a secondary school building in four distinctively different climate zones in North America (i.e., Toronto, ON, Canada; Vancouver, BC, Canada; Las Vegas, NV, USA and Miami, FL, USA). Soil moisture content, soil thermal properties, leaf area index, plant height, leaf albedo, thermal insulation thickness and soil thickness were used as design variables. Optimal parameters of green roofs were found to be functionally related to meteorological conditions in each city. In terms of energy savings, the results showed that the light-weight substrate had better thermal performance for the uninsulated green roof. Additionally, the recommended soil thickness and leaf area index for all four cities were 15 cm and 5 respectively. The optimal plant height for the cooling dominated climates is 30 cm and for the heating dominated cities is 10 cm. The plant albedo had the least impact on the energy consumption while it was effective in mitigating the heat island effect. Finally, unlike the cooling load, which was largely influenced by the substrate and vegetation, the heating load was considerably affected by the thermal insulation instead of green roof design parameters.

Keywords: green roof; energy performance; low impact development; heat island effect

---

1. Introduction

In recent years, due to a growing population and rapid urbanization, the demand for energy and water has increased significantly [1]. In 2010, buildings contributed to 32% of total energy used in the world and one-third of greenhouse gases [2,3]. Hence, buildings can play a significant role in limiting global warming and reducing climate change impacts. Green or vegetated roofs are regarded as an appropriate solution for the reduction of heating and cooling loads in buildings. Vegetated roofs mitigate the heat flux through the roof by shading the roof surface with plants, evapotranspiration of the plants, and the additional thermal insulation and mass, which are caused by the growing media [4]. There is a wide range of investigations that show the other advantages of green roofs such as reduction of stormwater runoff [5–8] as a Low Impact Development (LID) option, urban heat island effects [9–11], sound insulation [6,12], enhancement of the life span of a roof [13], creation habitats for species, and improving air quality by absorbing CO_2 and producing O_2 [14–20]. Due to the aforementioned benefits of green roofs, many local authorities in different countries have implemented the policy for using green roofs in new buildings. For instance, in Toronto, Canada, new buildings with a surface area of more than 2000 m^2 should have between 20% and 60% vegetation coverage of the roof area. In Tokyo, Japan newly constructed buildings must have at least 20% vegetated roof; in Portland, USA, all new buildings should have at least 70% green roof coverage of their rooftops; and in Basel, Switzerland, the roofs of new or retrofitted buildings need to be covered with at least 15% vegetation [1]. It is important to note that the performance of green roofs depends on different design variables such as growing media, composition and thickness, slope of roof, type of vegetation, and hydrologic variables such as rainfall and antecedent dry weather periods (ADWP). Additionally, the design variables will be different in different climate zones. Hence, these are the main obstacles for policymakers, designers and stormwater managers for implementing a green roof system in different climates [21].

Although the environmental advantages of green roofs is the main goal of local authorities in most cities, the thermal benefits of green roofs are of great importance and should be considered by the constructors and engineers for designing this type of roof.

A wide range of experimental and numerical studies to evaluate the effect of a green roof on the energy consumption of buildings in different climates have been conducted. For instance, different types of green vegetation such as sedum, grass and graminaceous plants have been used to analyze their impacts on the energy performance of a one-story office building in European climates. The results of energy modeling indicated that the cooling load consumption could be reduced between 1 to 11% in warm climates and up to 7% in cold climates [17]. Another study, evaluating the energy performance of an extensive green roof with Origanum heraclioticum plants for a supermarket in Athens, Greece showed that the cooling and heating loads were reduced by 18.7% and 11.4%, respectively [22]. Furthermore, different plant species with different height results in the literature are not consistent with one green roof study showing no effective energy savings of a mock up building in Pennsylvania, USA [23]. Another study in the Mediterranean climate in Italy compared the energy performance of a conventional roof with an extensive green roof covered with sedum plants, in which the green roof had a 100% reduction in thermal energy entering the roof in the summer and 30 to 37% of heat loss during the winter [24]. Jaffal et al. [4] conducted a study on the effect of a green roof on the energy performance of a single-family house with insulate roof in three different climates. The results of a simulation showed that heating increased by 8% in a Mediterranean climate while it did not have an effect on the heating load of the temperature climate. However, in cold climates (Stockholm), due to the insulation effect of a green roof, the heating load decreased by 8%. The results of energy modeling of a two-story nursery building with conditioned space showed that the cooling load reduced by 6–33%. The heating load increased by 3–9% for the insulated roof; however, for the uninsulated roof there was a reduction in heating demand [25].

To properly evaluate the influence of green roofs on the building energy performance, architects and engineers should evaluate the importance of design variables, external conditions and building type. The main design parameters that influence the thermal performance of green roofs are plant coverage and growing media. The main characteristics of plants that have to be considered in the design are leaf area index (LAI), plant height, stomatal resistance, leaf albedo and emissivity. Similarly, the growing media design characteristics are thickness, thermal properties and moisture content.

Several studies have analyzed the energy performance of green roofs by conducting a parametric analysis of one or more of these design variables. For instance, the energy consumption varied from 1% to 15% for a five-story building in Singapore by changing the soil thickness, moisture content of the substrate and plant type [26]. A parametric study in a Mediterranean climate showed that the LAI was the most influential parameter in reducing the cooling demand through evapotranspiration [27]. Furthermore, the LAI and substrate thermal properties were the most effective parameters in reducing cooling and heating loads for a supermarket in a semi-arid climate [28]. Sailor et al. [29] investigated the effect of climate, LAI and soil thickness in four USA cities including Houston, New York, Portland and Phoenix. The results reported that the green roof with deeper soil thickness could reduce the heating demand in cold cities. However, the LAI saved energy for the cooling dominated cities by using evapotranspiration and surface shading of plants. The investigation in Midwestern USA, for a city with hot and humid summers and cold and snowy winters, concluded that a green roof reduces the heat flux in summer and winter by 16% and 13%, respectively [30]. Ascione et al. [17] used the EnergyPlus®software to compare the effect of leaf index area on green roof performance in two different cities in Tenerife, Spain and Oslo, Norway. The investigation found that by increasing the leaf index area from 0.8 to 5 the energy savings for the cooling load in Tenerife increased from 1% to 11%, and the heating energy savings in Oslo was 5% with LAI = 5, and the savings was 6% with LAI = 0.8 (because of lower solar absorption). In another study [31] using EnergyPlus®in Portugal, the thermal performance of three types of green roofs (extensive, semi-intensive and intensive) with two conventional roofs with different surface color (black and white) was studied. Additionally, the effects of plant height and leaf area index were evaluated. The results showed that the semi-intensive and intensive green roofs with moderate thermal insulation had better energy saving compared to the black and white roofs. However, the extensive green roof had good performance in buildings without insulation (old buildings). In addition, the leaf area index and plant height improved the energy savings in cooling seasons considerably and the soil depth had a major effect in heating seasons. Design Builder®was used in a numerical simulation [32] to study the effects of leaf area index, soil depth and insulation thickness in including warm-dry, warm-coastal and cool-humid climates. The results demonstrated that the insulation thickness had a positive effect in reducing the energy use intensity (EUI) effect in Phoenix while it was reversed for Chicago. The EUI in Los Angeles was affected by the leaf area index. Gomes et al. [33] conducted a parametric analysis to analyze the impact of design parameters such as plant height and LAI on cooling and heating loads in a Mediterranean climate. The results showed that the cooling load increased up to 365% with a LAI of 1 compared to the LAI of 5. However, the heating load was influenced mostly by soil thickness, whereas heating load increased by 140% with a reduction of soil thickness from 1 to 0.1 m. Furthermore, with lower plant height (0.05 m) and LAI of 1 heating load reduced 23% and 18%, respectively.

The aforementioned studies were mostly for warm and Mediterranean climates and only a few research studies were conducted on the energy savings of a green roof in cold climates. The effects of snow depth and vegetation types were evaluated in Saint Mary's University, which is located in Halifax, Canada [34]. The research found that the green roof had lower heat loss when compared to the conventional roof. However, there was no difference in heat loss between the green roof and the conventional roof when the growing media was frozen or snow covered the roof surfaces. In addition, the variety of vegetation affected the depth of snow, duration of snow coverage and substrate temperatures. The effect of phase change in the green roof growing media was recently investigated during the winter at Purdue University [35]. The analysis showed that the green roof reduced heat loss up to 17.9% during the entire winter in comparison with the conventional roof. Additionally, during the phase change period (liquid to ice) the growing media stored a lot of energy and as a result, the growing media had a higher temperature. The heat loss was reduced during the phase change period in the winter by 19% while the heat loss reductions during the period without phase changes (the water was completely liquid or ice) were between 15–17%.

In spite of government funding for implementing the green roofs on buildings in Canada, few studies have considered the energy benefits of a green roof in this country, and these were mostly experimental. For example, results of a study in Toronto [36] for a single family and low-rise commercial buildings concluded that $11 Million of energy can be saved by using a combination of cool roofs and shaded trees. The study also reported that the urban heat island (UHI) effect of green roofs could reduce the electricity demand for cooling systems during the summer. It noted that a 0.6 °C increase in ambient temperature would increase the cooling peak demand from 1.5% to 2%. The National Research Council of Canada (NRCC) studied, analyzed and compared a typical extensive green roof with a conventional roof in Ottawa [37]. The results showed that the green roof reduced the heat flow through the roof in summer considerably more than in winter. The energy consumption for space conditioning was 75% lower than the reference roof. Additionally, during the days without snow, the green roof was slightly better than the conventional roof because of the growing media insulation effect. However, when the growing media was frozen, the thermal conductivity increased and the insulation effect was diminished. Therefore, both roofs had similar heat loss under the snow coverage. Another experimental study by NRCC [38] analyzed two different green roofs with a reference roof in

Toronto. The differences between green roofs were in the thickness and color of the growing media. The analysis illustrated that both green roofs reduced the heat gain and heat loss through the roof as well as peak load, significantly more than the reference roof. The green roofs reduced the heat gain during the summer by 70–90% and heat losses during the winter by 10–30%.

The British Columbia Institute of Technology (BCIT) conducted an experimental study to compare the energy performance of two green roofs with different growing media thicknesses with a reference roof on the west coast of British Columbia [39]. The study concluded that the green roof with thinner growing media had a smaller heat loss compared to the thicker growing media. Similar to the previous studies, the results showed that green roofs had a better performance compared to the reference roof and they saved energy through the roof during summer and spring from 83 to 85%, during the winter and fall from 40 to 44% and overall, a 66% reduction of annual energy consumption. A recent study for an office building in the Ryerson University in Toronto [40] analyzed the effects of LAI and soil thickness on energy consumption. The results showed that the growing media thickness was more effective in the reduction of energy. Additionally, the heating demand was reduced by 3% compared to the conventional roof.

One of the most important parameters that influences the effectiveness of a green roof is the thermal insulation of the roof: a green roof without thermal insulation reduces the cooling load considerably more than in the Mediterranean climate [27]. Another study in this climate demonstrated that the insulated green roof consumed more energy than the conventional roof [31]. A single-family residential building in the Oceanic climate of LaRochelle, France proved that uninsulated and insulated green roofs saved energy 48% and 10% more than the conventional roof, respectively [4]. Similarly, the energy reductions for well-insulated, moderately insulated and uninsulated green roofs for an office building in the Mediterranean climate of Athens (Greece) were 2%, 7% and 48%, respectively [41]. An experimental and simulation study in four U.S. climates concluded that the green roof with insulation was less effective in reducing energy consumption [42]. Pablo in his study proved that the green roof without insulation could have appropriate performance in the summer but not in the winter, as a lower heat transfer coefficient was needed for reducing the heating load [43].

Previous investigations mainly focused on the energy saving of a green roof in commercial or residential buildings and a couple of studies were conducted on other types of buildings such as retail stores or supermarkets [1,28]. School buildings in North America tend to be very large, and are designed with varying functionality (such as pools, gymnasiums and classrooms) and have significant energy consumption. However, the effect of vegetated roofs on the energy performance of this type of infrastructure has not been taken into consideration. The National Center for Education Studies reported that the number of buildings allocated for education in the USA was more than 100,000 [44]. It is notable that for each of the aforementioned schools, a considerable amount of energy was consumed to provide electricity for the students and teachers. According to a report by Xcel Energy [44], almost \$6 million per year is spent for the required energy in schools. According to a survey by the U.S. Department of Energy in 2013, school buildings consumed 8% of energy among commercial and institutional buildings [45]. Likewise, in Canada, schools are regarded as one of the major consumers of energy with an energy intensity of 0.88 GJ/m^2. According to Natural Resources Canada (NRCan), the average annual energy consumption of schools in Canada is 472 kWh/m^2 [46]. School buildings in Ontario make up a 30% proportion of energy consumption in the public sector [47]. Another study by NRCan showed that after office and food storage buildings, as well as schools are the third largest energy consumer in Canada with 8% energy consumption among commercial and institutional buildings [48].

Despite the necessity of energy savings in this sector in North America, previous research has mostly focused on specific locations and buildings when analyzing the effectiveness of green roofs on energy consumption. Furthermore, there has been no comprehensive study in this region to analyze the influence of green roof design parameters such as LAI, plant height, plant albedo, soil thickness, thermal insulation thickness and soil thermal properties, on the use of energy in buildings. In this research a parametric analysis focusing on secondary school buildings in different climates of North America was conducted in order to provide insight into the influence of these important design parameters on school building energy budgets. The selected climates included the cool-humid climate of Vancouver, Canada, the cold-continental climate of Toronto, Canada, the hot-dry climate of Las Vegas, USA and the hot-humid climate of Miami, USA.

2. Methodology

The methodology employed in this research involved simulating the energy balance of school buildings with two different green roof designs in the four climates. A parametric analysis of design parameters was conducted within this simulation in order to determine their influence on energy balance. This study focused on the influence of substrate (the vegetation growing media) thermal properties and moisture content on the energy performance of the green roof assembly by considering two different substrates: (1) heavy-weight and (2) light-weight based soil.

2.1. Simulation Tool

EnergyPlus, which is one of the most well-known energy modeling software available, was used in this study. Developed by the U.S. Department of Energy (DOE), it combines two energy modeling programs BLAST and DOE-2, which were developed by the U.S. Department of Defense and the DOE, respectively [49]. EnergyPlus calculates the required energy for heating, cooling, lighting, and ventilation as well as water heating in buildings, by considering all outdoor and indoor parameters including geographical location, outside temperature, solar radiation intensity and direction, wind speed and direction, glazing shading and radiation characteristics, which were provided in EnergyPlus Weather (EPW) format file, along with generated heat by internal loads such as occupant density and activities, lightening and miscellaneous equipment, infiltration, etc. The heat flux through the building envelopes was calculated based on the BLAST conduction transfer function (CTF) [49].

EnergyPlus 8.8 can model green roofs. This study used the green roof module developed by Portland State University, which was introduced in April 2007 and implemented in EnergyPlus [50]. The green roof model in EnergyPlus 8.8 allows the user to implement the green roof as an outer layer of roof assembly. The user can also vary different green roof design parameters such as growing media thickness and thermal properties, plant canopy density, plant height, stomatal resistance, and soil moisture conditions (including irrigation and precipitation) [49]. The energy balance in green roof assemblies is determined by balancing solar radiation (similar to that of a conventional roofing system). The solar radiation is balanced by the sensible and latent heat flux (evapotranspiration) through the soil and plants, conductive heat transfers through the soil, and long-wave radiation to and from the soil and plant surfaces. It should be noted that the rate of evapotranspiration is influenced by external conditions such as relative humidity, solar radiation, ambient air temperature, wind speed and substrate moisture content, which was considered when analyzing the results.

2.2. Building Model and Green Roof Parameters

EnergyPlus has a number of prototype buildings according to ANSI/ASHRAE/IES Standard 90.1–2013, and in this research the "secondary school" building was selected [51,52]. The building geometry was created using SketchUp. SketchUp has an OpenStudio plug-in that allows SketchUp building geometry to be used by EnergyPlus. The geometry of the building is shown in Figure 1. The school building consisted of two floors, 19,592 m^2 of conditioned space, 11,902 m^2 of roof area, a window to wall ratio of 33% and a total of 46 thermal zones, which were ventilated with a VAV system. The building had constant thermostat set points for both heating (21 °C) and cooling (24 °C), an occupancy of 0.3 people/m^2, and a lighting intensity of 12.5 W/m^2. A metal roof assembly with an albedo of 0.2 was considered for the building. The characteristics of the materials used in the roof structure were defined based on the EnergyPlus software library, as shown in Table 1. Schematics of the roof structure and green roof system are shown in Figure 2. The U-value of the walls and windows were 0.5 W/m² K and 3.2 W/m² K, respectively.

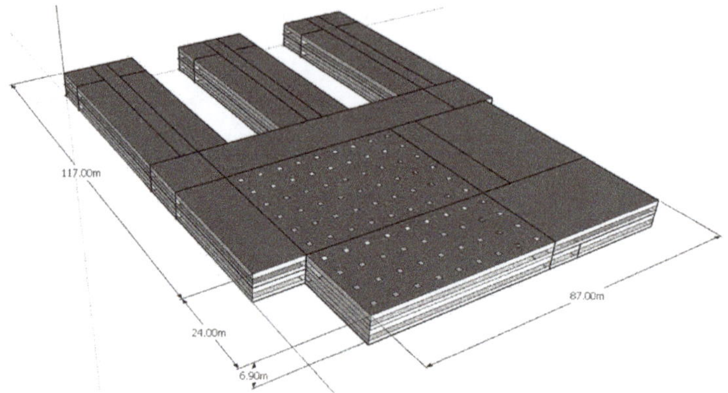

Figure 1. Secondary school building geometry.

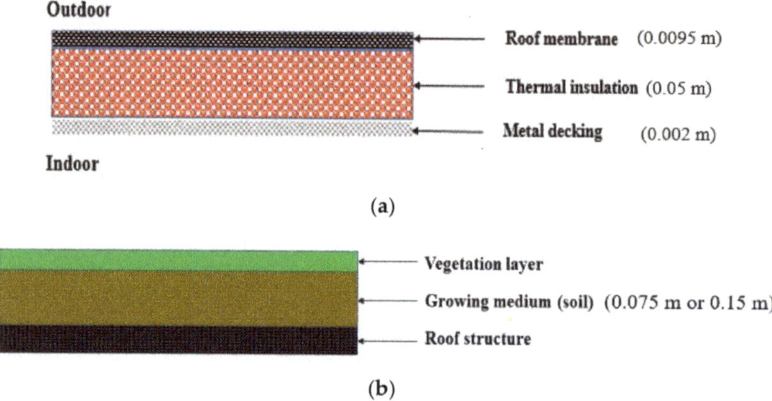

Figure 2. (a) Schematic of roof structure; (b) schematic of green roof system.

Table 1. Thickness and thermal properties of the roof materials layer [53].

Material	Thickness (m)	Density (kg/m³)	Thermal Conductivity (W/m·K)	Specific Heat J/(kg·K)
Roof membrane	0.0095	1121	0.16	1460
Thermal insulation (polyurethane)	0.05	40	0.04	1600
Metal decking	0.002	7680	45	418

To design and optimize the thermal performance of building envelopes, such as walls and roofs, the climatological conditions of buildings should be considered by designers. The heating degree days (HDD) and cooling degree days (CDD) are measures of heating and cooling loads in buildings that are often used. ASHRAE climate zones is divided based on two parameters: temperature and moisture. The temperature is shown with a number ranges from 1 (hottest) to 8 (coldest), and moisture is demonstrated with a letter including A (moist), B (dry) and C (Marine). Hence, to study the effect of different climates on the energy performance of a green roof, the same building was considered in four different climates consisting of hot-dry, hot-humid, cool-humid and humid-continental. Here, simulations were performed for each city using third typical meteorological year (TMY3) weather files for U.S. cities [54] and Canadian Weather for Energy Calculations (CWEC) weather files for Canadian cities [55]. The climatic information for the four selected cities is summarized in Table 2.

Table 2. Summary climate data for the four cities modeled.

City	Annual HDD Base 18 °C	Annual CDD Base 10 °C	Summer Conditions	Winter Conditions	ASHRAE Climate Zone *
Toronto	3956	1316	Humid, Warm	Cold	6A
Vancouver	2932	951	Moderate	Mild	5C
Las Vegas	1169	3908	Dry, Hot	Mild	2B
Miami	72	5447	Humid, Hot	Moderate	1A

* A indicates moist, B indicates dry and C indicates marine moisture conditions.

In order to understand the impacts of different green roof parameters on the energy performance of buildings, with the exception of the roof assembly, all building characteristics such as building envelopes, internal loads and schedules remained unchanged. For comparison, the thermal responses of two conventional roofs with (R-7) and without insulation were also studied. An uninsulated extensive green roof with shallow growing media and an adequate structure for building retrofitting was considered in this study. Additionally, green roof irrigation was not considered in this investigation. The vegetation types, which are used in green roofs must be adapted to harsh environmental conditions on the roof. Vegetation type could change the hydrologic behavior of green roofs in terms of interception, uptake, transpiration and storage of water [21]. For instance, grass has better performance than sedum in storm water retention while sedum is one of the most compatible plants used in extensive green roofs [21]. Hence, in selecting the vegetation type, the hydrologic behavior of the green roof along with its effect on energy performance had to be considered. Based on plant type, the design parameters such as LAI, plant height and leaf albedo varied. For instance, the LAI of sedum typically ranges from 2 to 4 [56]. Feng et al. [57] reported a LAI of 4.6 for Sedum lineare. Furthermore, [17] used sedum short and tall cut with LAI of 0.8 and 3, respectively. Graminaceous short cut and tall cut has a LAI of 2.5 and 5, respectively [17]. Gagliano et al. [58] used a combination of different plants such as sedum, mosses, graminaceous or other succulent type plants with a LAI of 5, as the vegetation layer for an extensive green roof. The plant height depends on the type of plant and the type of green roof. For instance, the plant height of sedum sediforme with flowering stems range from 25 to 60 cm [59]. Sedum short cut and long cut have plants of heights of 10 cm and 30 cm, respectively [17]. Likewise, different plants have different plant albedo. For example, in [38] plant albedo of different types of sedum were studied where sedum tomentosum and mixed sedum species had a plant albedo of 0.23 and 11, respectively. Additionally, Ascione et al. [17] considered a plant albedo of 0.22 for both short and tall cut sedum for modeling green roofs.

In this study, parameters were selected for an extensive green roof and consisted of different growing media, leaf area index (LAI), plant height, plant albedo, soil thickness and thermal insulation thickness, as shown in Table 3. Substrate and vegetation plant properties were selected based on previous studies [1,17,33,57–59]. Light-weight and heavy-weight substrates were not input parameters in EnergyPlus; however, their properties were considered as inputs in the software. It is important to note that the assumptions that were made for the parametric studies do not represent the realistic conditions of green roofs, such as a constant height or LAI during the year. However, these assumptions were necessary to conduct critical and meaningful parametric analyses. The constant characteristics of the green roof are shown in Table 4.

There are two methods of moisture diffusion for the substrate in EnergyPlus: (1) Simple, and (2) Advanced. The Simple model is the basic model which is based on the moisture diffusion between two layers in the soil. At each time step, the soil properties are updated and based on the moisture content of the layers whereby the moisture distributes at a constant rate from a higher moisture layer to the lower moisture layer. In the Advanced method, moisture diffusion is based on the Schaap and van Genuchten model [60]. This model divides the soil into different nodes and then distributes the moisture through the layers. In this study we used the Advanced model because the output it produced in terms of moisture content changed in the soil and was more realistic in comparison to the Simple model [61]. The influence of substrate thermal properties and moisture content on the energy performance of the green roof assembly was studied by using two different substrates: (1) heavy-weight and (2) light-weight based soil [62]. The light-weight growing media had a lower density and thermal conductivity in comparison to the heavy-weight growing media. It should be noted that in EnergyPlus, the thermal properties of soil are constant, as set out in the input file, and do not vary with moisture content [63]. In addition to this, the energy balance of snow and the effect of the freezing of soil on heat transfer are not considered in the current versions of this software [64]. However, it allows users to provide input for precipitation and irrigation schedules, and track the variation of soil moisture content both daily and monthly. The soil properties used in this study were taken from the literature [62]. The materials used in the light-weight soil were 75% pumice, 10% compost and 15% sand. This soil had the highest moisture capacity, lowest thermal conductivity, and lowest density among all soils. The heavy-weight soil consisted of 50% expanded shale and 50% sand, and had the highest density and thermal conductivity among the soils. To evaluate the effect of soil moisture content on the energy performance of the green roof, substrate thermal properties were obtained for three levels of moisture content (20%, 60% and 100% saturation) as shown in Tables 5 and 6. These properties were considered as constant inputs for the software while the moisture content of soil varied throughout the year in the software.

Table 3. Design parameters for the parametric analysis.

Parameter	Value
Growing Media	Light-weight and heavy-weight substrates
Leaf Area Index (LAI)	0.1, 2, 5
Plant Height (cm)	10, 30
Leaf Reflectivity (Albedo)	0.11, 0.23
Soil thickness (cm)	7.5, 15

Table 4. Green roof constant parameter values.

Parameter	Value
Leaf Emissivity	0.95
Thermal Absorptance	0.9
Visible Absorptance	0.75
Minimum Stomatal Resistance	300 (s/m)
Roughness	Medium Rough
Saturation Volumetric Moisture Content of the Soil Layer	0.5 (m^3/m^3)
Residual Volumetric Moisture Content of the Soil Layer	0.01 (m^3/m^3)
Initial Volumetric Moisture Content of the Soil Layer	0.2 (m^3/m^3)
Moisture Diffusion Calculation Method	Advanced

3. Results and Discussion

3.1. Influence of Soil Type and Moisture Content

As can be seen in Tables 5 and 6 the thermal conductivity and density increased with the increasing moisture content of the soil. Additionally, as the pores of the soil filled with water, the specific heat capacity of the soil increased because of the fact that water has a higher thermal storage capacity than air. However, by increasing the water content, the soil reflectivity (albedo) reduced significantly as the soil surface was wet and absorbed a higher fraction of solar radiation compared to the dried surface.

Thermal diffusivity, which represents the ability of materials to conduct energy relative to its thermal capacity plays a key role in evaluating the thermal response of materials. The effect of thermal diffusivity of soil becomes more important during the summer where the transient heat transfer is more significant because of higher temperature oscillations compared to the winter.

Table 5. Light-weight substrate thermal properties for different levels of moisture [62].

Saturation Level (%)	Density (kg/m^3)	Thermal Conductivity W/(m·K)	Specific Heat J/(kg·K)	Albedo	Thermal Diffusivity (m^2 s^{-1})
20	765	0.21	1284	0.27	2.12×10^{-7}
60	870	0.31	1602	0.18	2.22×10^{-7}
100	934	0.41	1853	0.12	2.36×10^{-7}

Table 6. Heavy-weight substrate thermal properties for different levels of moisture [62].

Saturation Level (%)	Density (kg/m^3)	Thermal Conductivity W/(m·K)	Specific Heat J/(kg·K)	Albedo	Thermal Diffusivity (m^2 s^{-1})
20	1385	0.37	936	0.09	2.85×10^{-7}
60	1450	0.60	1035	0.04	3.91×10^{-7}
100	1500	0.84	1095	0.02	5.10×10^{-7}

The results in Tables 7 and 8 were for a green roof with a LAI of 2, plant albedo of 0.11, 30 cm plant height and soil thickness of 7.5 cm. The light-weight soil, because of high specific heat capacity and low thermal conductivity, had lower thermal diffusivity compared to the heavy-weight soil. While, it was expected that the heavy-weight soil would have better performance compared to the light soil during the summer months because of its higher thermal capacity, due to the low thermal diffusivity of the light soil, it was found to perform better during the large daily temperature fluctuations between day and night in the summer. This is illustrated in Table 7.

Table 7. Cooling load (kWh/m^2) variation with moisture content for insulated (I) and uninsulated (U) roofs in different climates for light-weight (LW) and heavy-weight (HW) soils.

City	Insulation	20% Saturated		60% Saturated		100% Saturated	
		LW	HW	LW	HW	LW	HW
Toronto	I	8.31	8.46	8.37	8.53	8.41	8.68
	U	11.11	11.48	11.85	12.42	12.37	12.90
Vancouver	I	3.01	3.08	3.03	3.13	3.05	3.15
	U	3.98	4.24	4.36	5.11	4.62	5.54
Las Vegas	I	41.81	45.52	42.13	42.95	42.30	43.52
	U	57.10	58.66	57.76	62.63	58.50	66.22
Miami	I	69.12	70.16	69.64	70.42	70.44	70.64
	U	82.93	84.24	84.81	89.37	87.22	93.70

It was also seen that by increasing the moisture content of the soils, the thermal diffusivity of light soil, unlike heavy-weight soil, remained almost unchanged. However, for the heavy-weight soil, by increasing the moisture content the thermal diffusivity increased. As a result, the cooling load (Table 7) for the uninsulated green roof with the heavy-weight substrate increased considerably more than the light-weight soil. Moreover, as can be seen in Tables 6 and 7, at a moisture level of 20%, the heavy-weight and light-weight soils had similar thermal diffusivity. Therefore, the cooling loads were almost equal to each other in this condition. However, at higher levels of moisture content, the light soil had a better thermal performance compared to the heavy-weight soil.

On the other hand, in winter, thermal conductivity plays a key role in heat transfer through the roof. The thermal conductivity of heavy-weight soil at a higher moisture level was almost twice that of light-weight soil (Table 5). Quite naturally, simulation outputs in Table 8 demonstrate that the heating load for the roof with light-weight soil was considerably lower than the same with heavy-weight soil. The importance of thermal conductivity on the heating load is more evident for cities such as Toronto, Vancouver and Las Vegas (cities that experience a cold winter). However, in Miami which has a mild winter, the heating load difference between the light-weight and heavy-weight soil was minimal (Table 8). That is to say that the heating load was not largely affected by the soil type in regions with this climate. Therefore, the uninsulated green roof with light-weight soil had better thermal performance in all four cities.

Table 8. Heating load (kWh/m^2) variation with moisture content for insulated and uninsulated roofs in different climates.

City	Insulation	20% Saturated		60% Saturated		100% Saturated	
		LW	HW	LW	HW	LW	HW
Toronto	I	142.60	145.23	144.23	146.46	144.90	147.00
	U	194.84	221.36	212.83	238.87	224.55	247.27
Vancouver	I	92.93	94.43	93.83	95.16	94.23	95.44
	U	125.10	143.24	137.00	154.93	144.56	160.83
Las Vegas	I	26.64	26.92	26.77	27.10	26.84	27.11
	U	35.72	40.74	38.83	45.10	40.82	46.84
Miami	I	2.25	2.26	2.25	2.29	2.26	2.31
	U	2.90	3.41	3.14	3.90	3.32	4.33

The results in Tables 7 and 8 show that the energy consumption of the building with an insulated green-roof assembly was not largely dependent on the soil type (i.e., light-weight or heavy-weight). These tables also show that the moisture content of the soil did not have an effect on the variation of cooling and heating loads when the roof was insulated.

For the parametric analysis conducted thereafter, the light-weight soil which consisted of 75% pumice, 10% compost and 15% sand, and with a 50% saturation level was considered. The thermal properties of the soil at 50% saturation for the light-weight soil included a density of 840 kg/m^3, a thermal conductivity of 0.28 W/m·K, a specific heat of 1500 J/(kg·K), an albedo of 0.2 and a thermal diffusivity of 2.85×10^{-7} m^2·s^{-1}.

3.2. Influence of Leaf Area Index (LAI)

Leaf Area Index (LAI) is defined as the projected area of all leaves divided by the soil surface area, which typically ranges from 0.001 to 5 [65]. The larger LAI indicates that less of the soil is exposed to solar radiation and as a result the soil surface temperature is lower. In this section an uninsulated green roof with a soil thickness of 7.5 cm, plant albedo of 0.11 and plant height of 30 cm was considered. Figure 3 shows the temperature variation of a green roof soil, in Vancouver, BC, with different LAI and outdoor air for a typical day in summer and winter. On a typical cold winter day, the soil temperature was warmer than the outdoor air temperature by an average of 6 °C, 5.7 °C and 5.1 °C when the LAI was 0.1, 2 and 5, respectively. This was attributed to the thermal storage behavior of the soil, i.e., heat is absorbed into the soil as it transfers from the warm interior to the cold exterior. It should be noted that with LAI equal to 5, the soil surface was still warmer than ambient air which indicates that the insulating behavior of foliage limits the heat losses from the soil surface to the ambient air. On a typical hot summer day, due to the higher rate of solar radiation absorption, the surface temperature was warmer than the outdoor air temperature by an average of 18 °C, 12 °C and 7 °C for LAI ranging from 0.1, 2 and 5, respectively. It was clear that the soil surface was shaded with higher plant coverage resulting in a lower surface temperature. Similar results and observations were made for the other cities too [61].

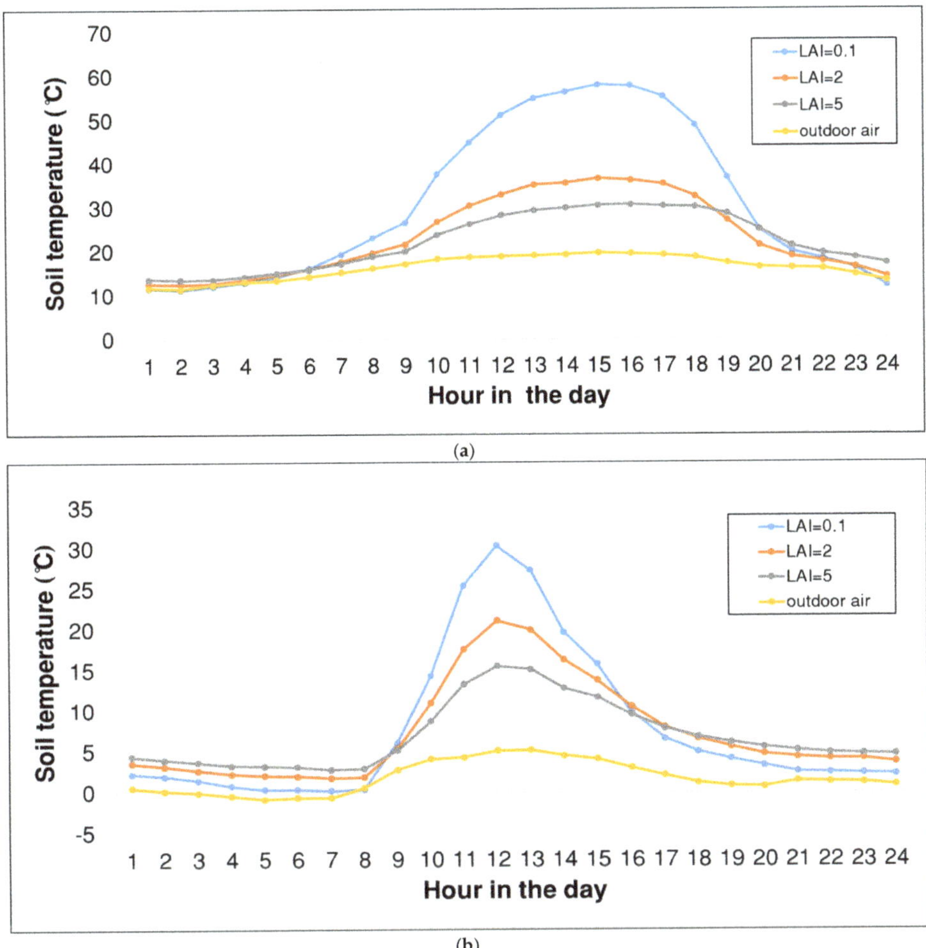

Figure 3. Hourly variations of soil surface temperature with different leaf area index (LAI) for a typical day in (**a**) summer and (**b**) winter for Vancouver.

The LAI has a major effect on the rate of evapotranspiration in green roofs. By increasing the LAI, the rate of evapotranspiration increases as the stomatal resistance of the plants exhibits an inverse relation with LAI. Therefore, it can be concluded that a greater LAI leads to a decrease in cooling load. Equation (1) shows the relationship between the LAI and stomatal resistance [50].

$$r_s = \frac{r_{s,min}}{LAI} f_1 f_2 f_3 \tag{1}$$

The actual stomatal resistance varies with the minimum stomatal resistance ($r_{s,min}$) at any time and it has an inverse relationship with LAI. To modify the stomatal resistance equation, fractional multiplying factors f_1, f_2 and f_3 which functionally relate to the incoming solar radiation and atmospheric moisture, were considered. It should be noted that the rate of evapotranspiration indicates the effectiveness of LAI. However, the plant evapotranspiration individually depends on external conditions such as solar radiation, wind speed, ambient air temperature, relative humidity and soil moisture content. Therefore, to compare the effectiveness of LAI on the building energy consumption for the four selected cities, the impacts of the aforementioned variables had to be considered in the analysis.

Because of the difficulty in measurement of the evapotranspiration process, the rate (or flux) of evaporation was defined to indicate the degree of evapotranspiration [66]. Figure 4 compares the average monthly evapotranspiration rate for all four selected cities at two different levels of LAI (5 and 2). It was evident that this rate during the winter was considerably lower than summer because of lower solar radiation and ambient temperature. Additionally, the rate of evapotranspiration was significantly lower when LAI = 2, because of less solar radiation absorption by the plants. The heating and cooling load variations with different LAI are shown in Figure 5. It was seen that in Las Vegas higher solar radiation intensity and ambient air temperature as well as lower air relative humidity contributed to the high rate of evapotranspiration and cooling load reduction among all four cities. However, in Miami because of high relative humidity, a low rate of solar radiation and wind velocity [61], the evapotranspiration rate remained almost unchanged even with greater LAI and the cooling load reduced slightly in comparison with other cities. The cooling load reduction in Miami was just 29% in comparison with 54%, 45% and 41% reduction in Vancouver, Toronto and Las Vegas respectively as shown in Figure 5. This indicated the effect of solar radiation, wind speed and relative humidity on the effectiveness of LAI and evapotranspiration rate in different climates. The rate of evapotranspiration in Vancouver was higher than that of Toronto during the cooling seasons because Vancouver has lower relative humidity and higher solar insolation than Toronto [61].

As mentioned previously (Figure 3) the soil surface temperature decreased with increasing LAI. The reduction of soil temperature contributed to the higher heat losses through conductive heat transfer in the soil during the heating seasons. The heating load in Las Vegas and Miami increased by almost 21% and 6.5%, respectively (Figure 5). It should be noted that this trend for Vancouver and Toronto was reversed. This suggested that during the winter the rate of evapotranspiration in these two cities was almost zero compared to that of Las Vegas and Miami because of lower ambient temperatures and solar radiation. In other words, during the winter, Vancouver and Toronto have many cloudy days and the number of sunny days are considerably lower than Miami and Las Vegas [61]. Therefore, greater plant coverage acts as a barrier for heat losses during the winter. The heat loss was reduced by almost 8% in both cities. It can be seen in Figures 3–5 that although the LAI had major impacts on the reduction of a long wave and short wave radiation incident on the roofing system during the summer, it influenced the heat losses and heat gains during the winter as well. However, the performance predominantly depends on the climatic conditions. Therefore, LAI plays a key role in decreasing or increasing the heat flux through the green roof. Additionally, in Figure 5 the heating and cooling loads of conventional roofs were compared with green roofs. The insulated conventional roof was insulated with 5 cm polyurethane. It was seen that the energy usage of building with uninsulated conventional roof was considerably higher than green roofs. It is important to note that the LAI of plants varies throughout the year, and would affect the heating and cooling loads. The results showed that heating and cooling loads variation with change of LAI mainly depended on climate zone. For instance, during the winter where lower LAI (LAI = 2) is expected, heating loads for Toronto and Vancouver were higher compared to the LAI = 5 while for Las Vegas and Miami is lower. Furthermore, it is worthwhile to note that the variety of vegetation affects the depth of snow, duration of snow coverage and substrate temperatures, in particular in cold climates like Toronto. However, in the EnergyPlus 8.8 the effect of freezing of the soil and snow cover in energy modeling were not considered.

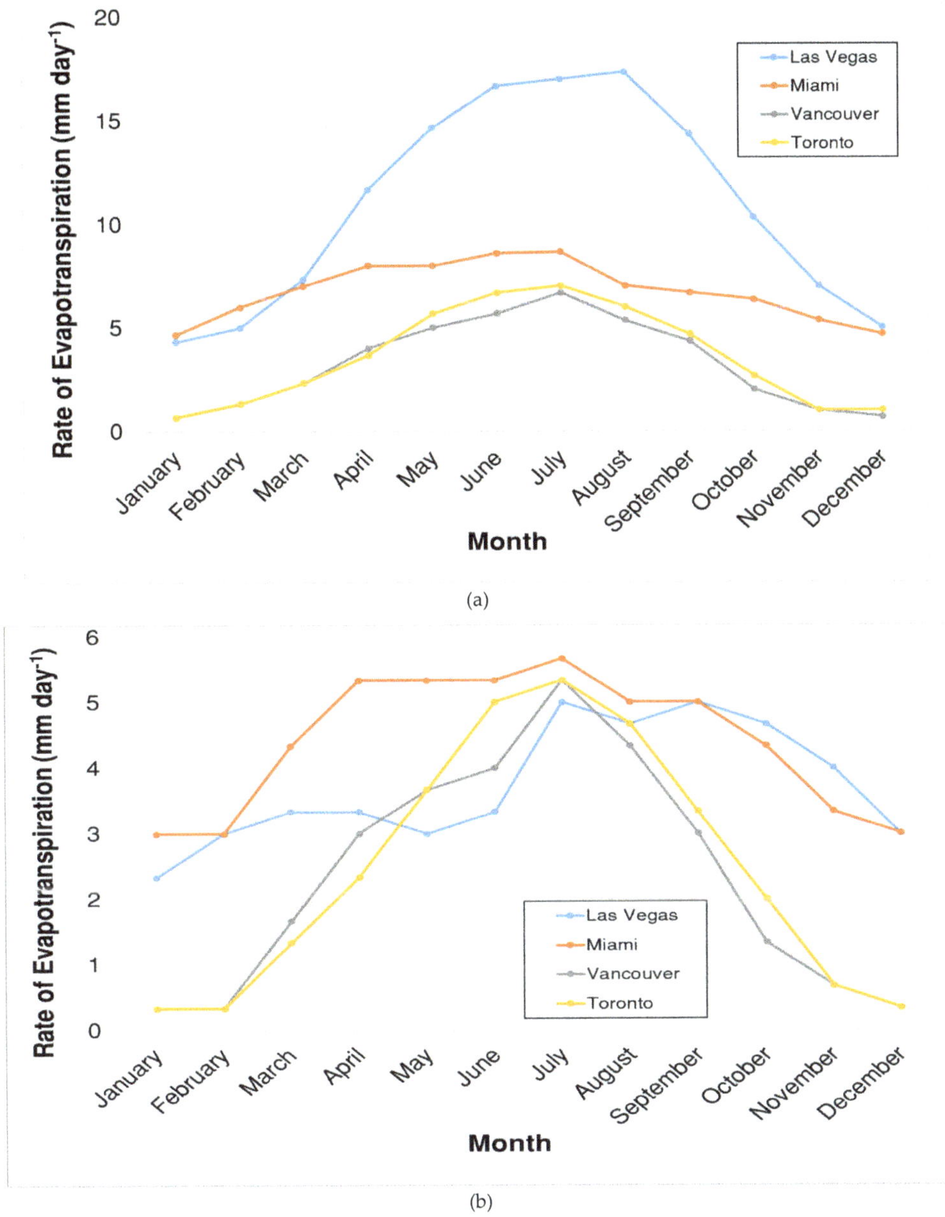

Figure 4. Monthly average rates of evapotranspiration for the (**a**) higher plant coverage (LAI = 5) condition and (**b**) for the lower plant coverage (LAI = 2) condition.

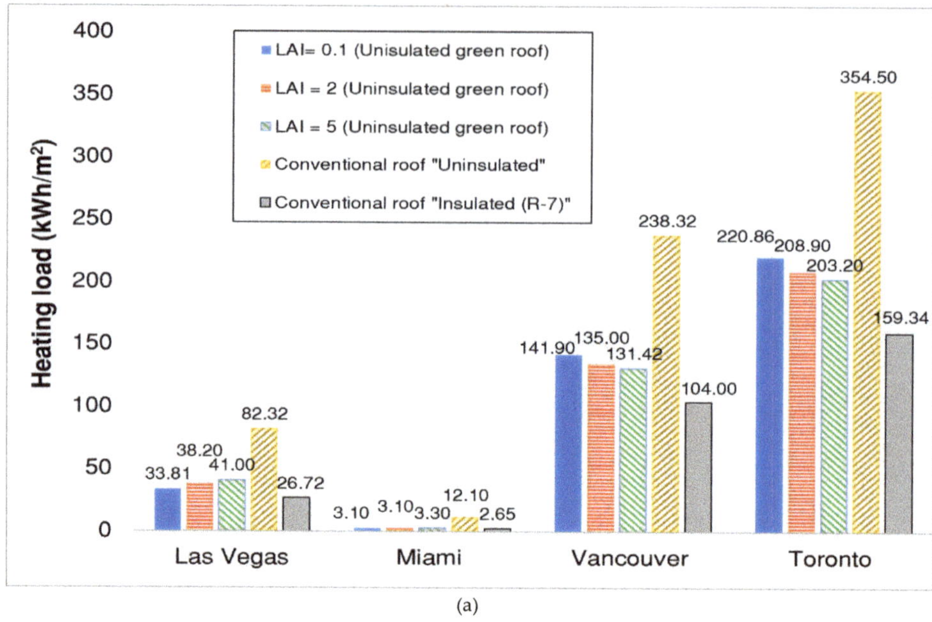

(a)

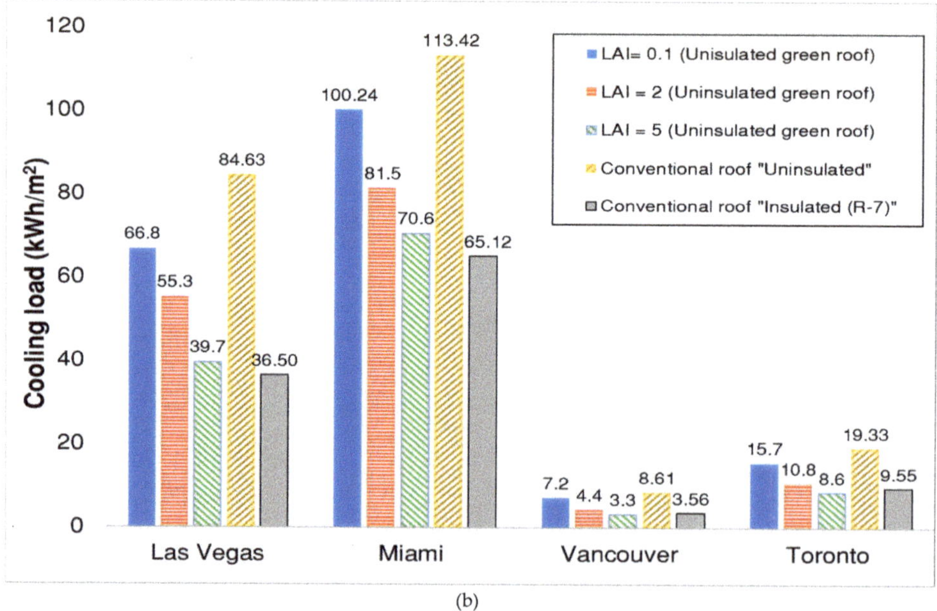

(b)

Figure 5. Sensitivity of the (**a**) heating load and (**b**) cooling load with LAI in different climates.

3.3. Influence of Soil Thickness

Increasing soil thickness results in lower heat flux because of thermal storage and insulation effects of the soil. The thicker soil contributes to a reduction in the operating costs of the HVAC system, however, it increases the initial costs [65]. In this section an uninsulated green roof with soil thickness of plant albedo of 0.11 and plant height of 30 cm was considered. It can be seen in Figure 6 that higher soil thickness produced higher reduction of heating and cooling load in all locations. The results in Figure 6b demonstrated that the highest percentage reduction in cooling load was in Vancouver, which was 27% when the soil thickness was 15 cm. It can be deduced that, as the summer in Vancouver is cooler compared to the other cities, the thicker soil has greater impacts on the cooling load. The minimum percentage reduction of cooling load was in Miami (6%) as it had a hot and humid summer and the extra thermal storage of the soil was not as effective as in other cities because of the low temperature fluctuation during the day. These reductions were lower when the LAI was equal to 5, which was justifiable because of the cooling effect of the plants. Similarly, as can been seen in Figure 6a during the winter, the soil thickness was more effective when the LAI was low. For instance, with LAI = 2, the heat loss was reduced by 30% in Miami when the soil thickness was doubled. However, this reduction in Toronto was found to be lower (16%) as a result of the much colder winters. With LAI equal to 5 the radiative heating capability of the thicker soil is reduced as the vegetation cools the roof and increases the heat losses. It is evident from Figure 6 that the soil with a 15 cm thickness saved energy in all climates. Similar to Figure 5, the heating and cooling loads of conventional roofs were compared with green roofs in Figure 6. It can be seen that the energy usage of a building with an uninsulated conventional roof was considerably higher than green roofs. However, an insulated conventional roof had better performance than a green roof where it was more evident in heating dominated climates. It is important to note that the effect of freezing on the soil on the energy simulation of a green roof in cold climates should be considered. To illustrate, with the frozen soil, the thermal conductivity of soil increased and the insulation effect of growing media would be diminished in this case, which would lead to higher heat losses through the roof.

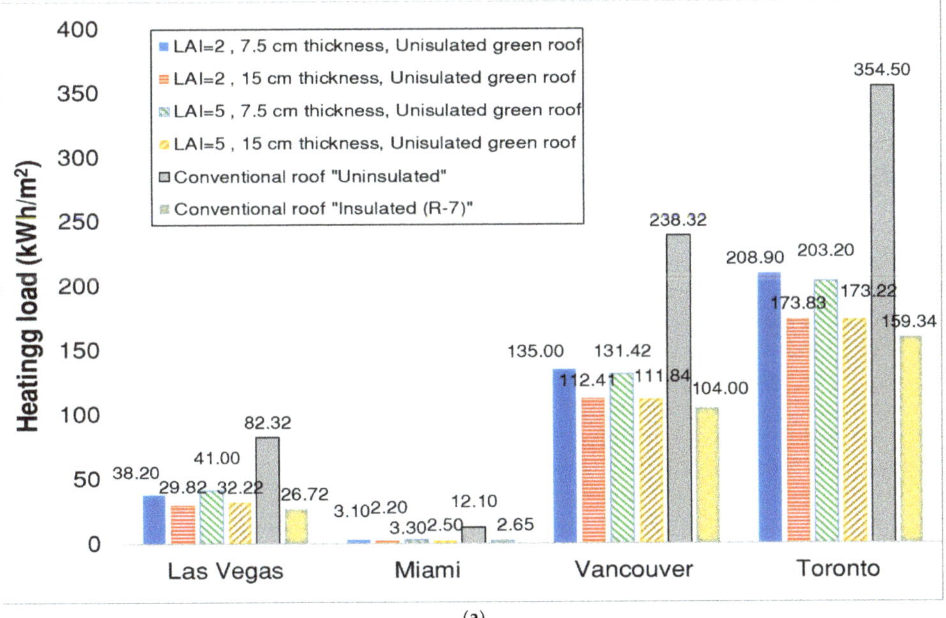

(a)

Figure 6. *Cont.*

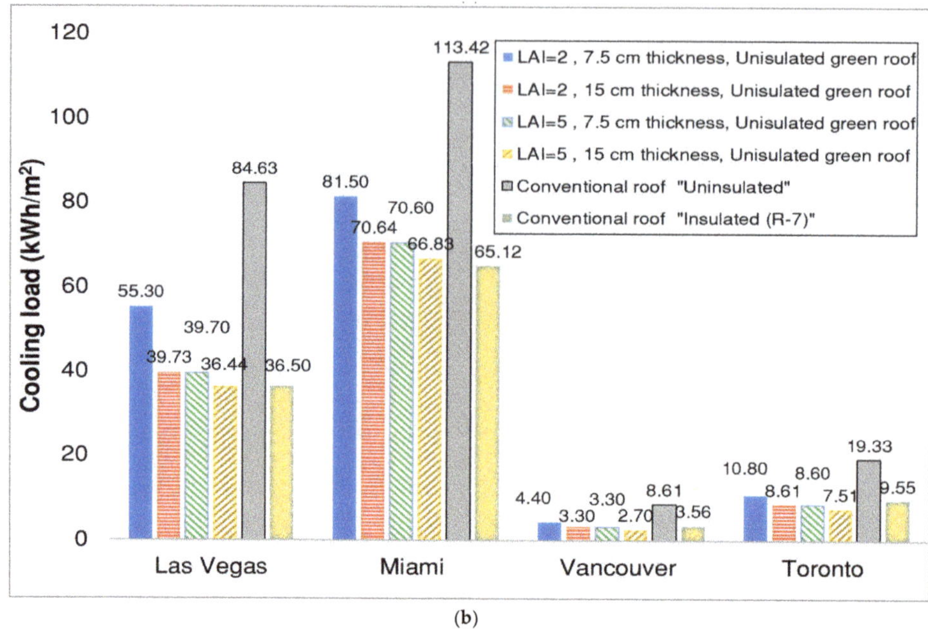

(b)

Figure 6. Sensitivity of (**a**) heating load and (**b**) cooling load with LAI and soil thickness in different climates.

3.4. Influence of Plant Albedo

The plant surface temperature had major impacts on the energy balance of a green roof [10]. This was due to the convective heat transfer between the leaves surface and air within the canopy. The higher foliage surface temperature results in increases in the ambient air and reduces the mitigating effects. The surface temperature depends on the amount of solar and thermal radiation absorption, emission, convective heat transfers with ambient air and the latent heat flux by transpiration [10]. The variations of vegetation latent heat fluxes, soil temperature and foliage temperature with plant albedo are presented in Figure 7 for a green roof with LAI = 5 and 15 cm soil thickness in Vancouver, BC. In this section, an uninsulated green roof with a soil thickness of 15 cm and LAI of 5 was considered for analysis. The results are available for other cities too [61]. It was observed that the plant with albedo 0.11 had higher foliage and soil temperature due to the higher solar absorption. The higher foliage temperature also contributed to the higher rate of transpiration. Nevertheless, the cooling effect of high transpiration (higher evaporation) was canceled out by the higher surface temperature (higher conduction heat transfer) of the soil. Similar observations are available for other cities too [61].

Similar to the other parameters, the effect of leaf reflectivity on the thermal performance of the green roof was evaluated. The results in Figure 8 show that the plant albedo did not have effect on the energy performance of the green roof because of the tradeoff between the cooling effect of evapotranspiration and the increase in soil temperature. The cooling and heating load in all four cities changed less than 1% by altering the plant albedo from 0.11 to 0.23. Therefore, although the plant albedo had a negligible effect on the green roof energy performance, for the cities such as Vancouver and Toronto which are heating dominated climates, the plant with lower reflectivity 0.11 was considered and for the Miami and Las Vegas which are cooling dominated climates the plant with higher reflectivity 0.23 was suitable. It should be noted that the variation of plant color throughout the leaf affected the leaf albedo. However, unlike LAI, from the results it can be concluded that variation plant color during different seasons has a very small effect on the heating and cooling variations.

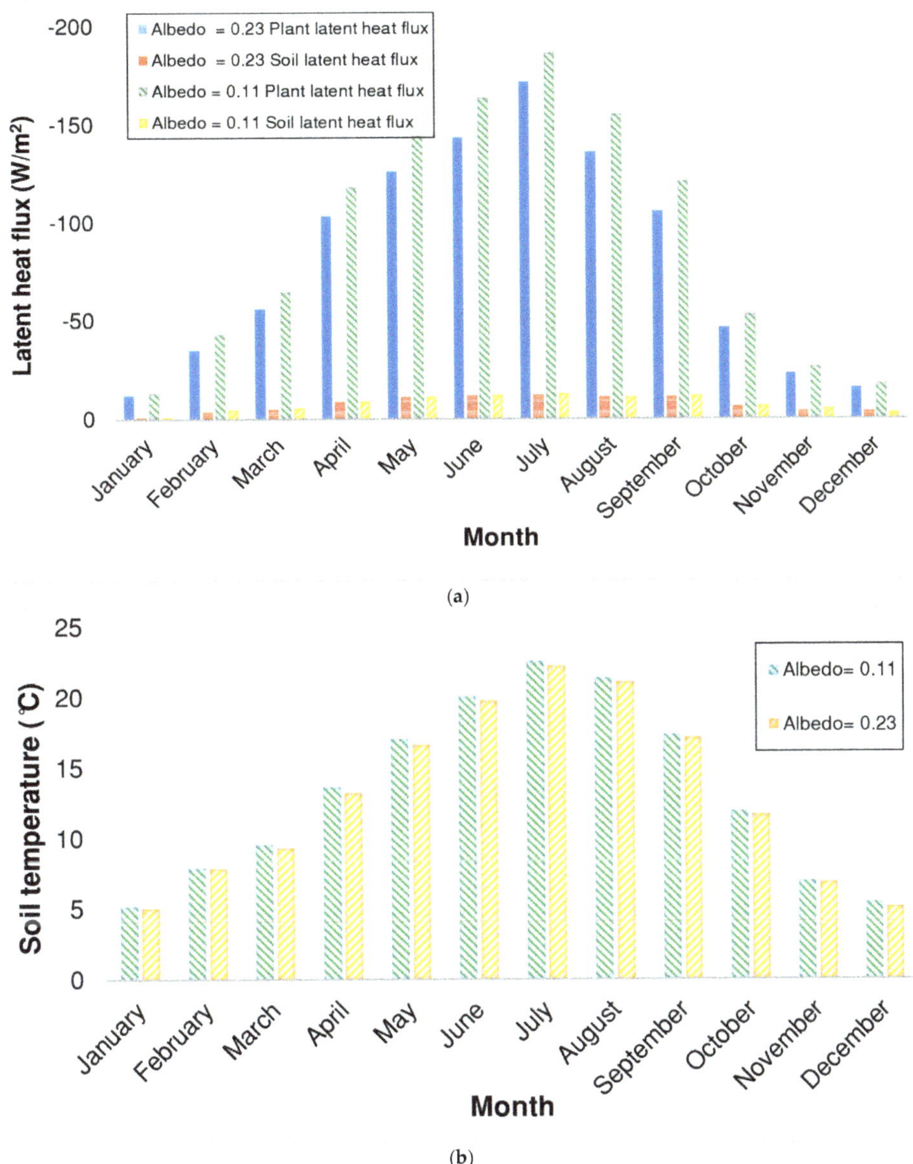

Figure 7. *Cont.*

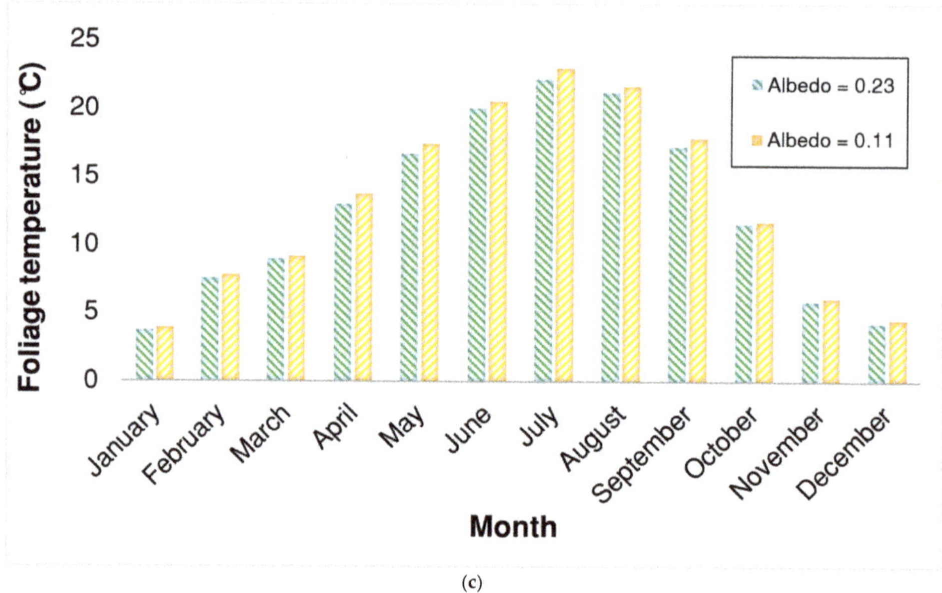

(c)

Figure 7. (**a**) Effect of plant albedo on the plant and soil latent heat fluxes; (**b**) monthly average variations in soil temperature; (**c**) monthly average variations of foliage temperature with plant albedo, in Vancouver, BC.

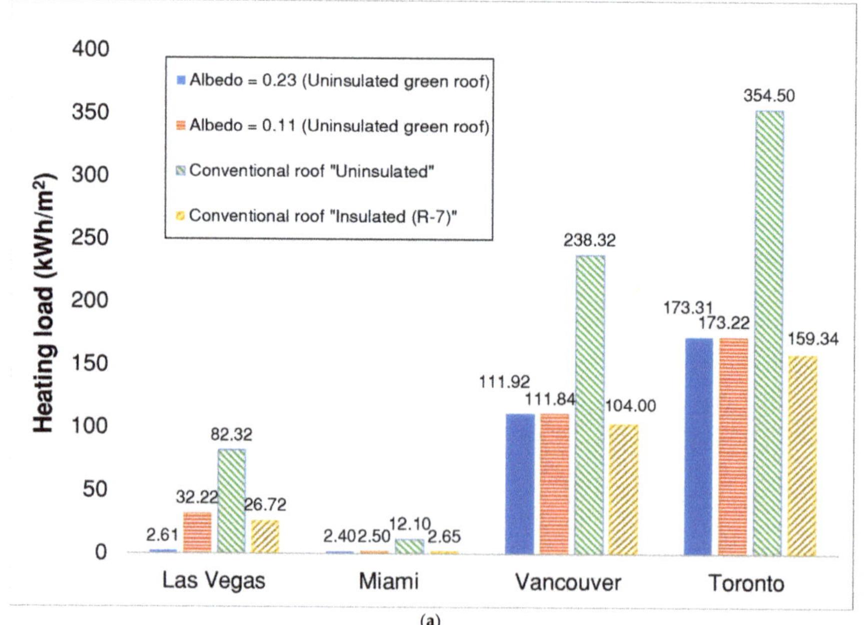

(a)

Figure 8. *Cont.*

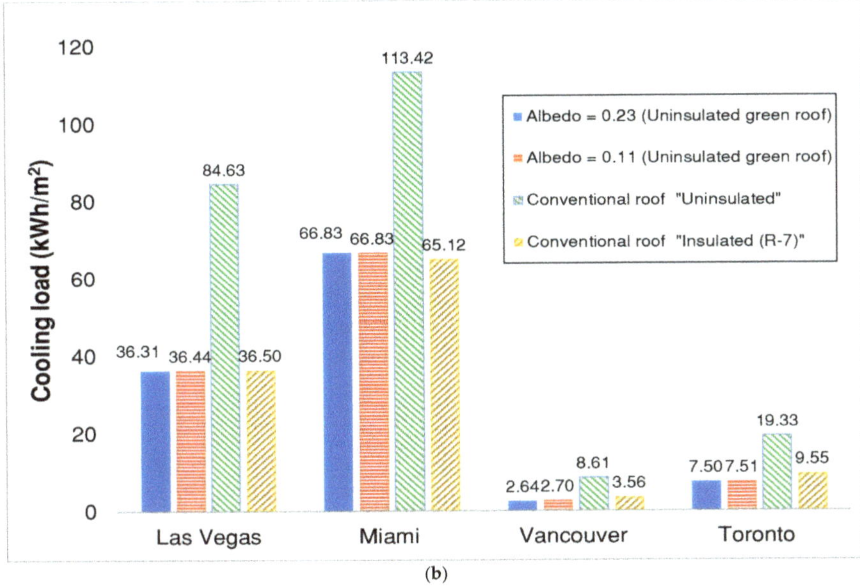

Figure 8. Sensitivity of (**a**) heating load, and (**b**) cooling load, with plant albedo in different climates.

3.5. Influence of Plant Height

The plant height, like LAI has impacts on the shading of the soil surface and the rate of evapotranspiration. For instance, with short plants the soil temperature increases more than with tall plants because a larger part of the soil is exposed to the solar radiation. Consequently, the solar radiation absorption by the soil cancels out the cooling impacts of the evapotranspiration process [65]. On the other hand, it is expected that increased plant height will reduce cooling load. With increased plant height, the wind velocity within the canopy increases, leading to higher rates of evapotranspiration. This is strongly related to the formed thick boundary layers on the leaf surface where air close to the leaf surface has marginal movement. Water vapor which leaves the stomata has to transfer through this motionless boundary layer to reach the atmosphere and, hence, the rate of transpiration is slower when the boundary layer is thicker or in other words, when the wind velocity is low.

The aerodynamic resistance factor shows the resistance that is caused by the boundary layer for exchanging moisture. Wind velocity, surface roughness and stability of the atmosphere are the main parameters which influence the aerodynamic resistance. The aerodynamic resistance of a plant is inversely proportional to wind velocity W_a [50]. Aerodynamic resistance also affects the leaf surface wetness, such that as the aerodynamic resistance approaches zero, the moisture is transferred rapidly to the leaf surface and evaporated easily [50].

In this section an uninsulated green roof with a soil thickness of 15 cm, plant albedo of 0.11 for cold climates and 0.23 for warm climate, and LAI of 5 was considered. The effect of plant height and the LAI interaction on heating and cooling loads are shown in Figure 9. It can be seen that the growth of the plant does not always lead to a cooling load reduction and the effectiveness of plant height depends on LAI and climate zone. For instance, there was an interactive relationship between LAI and plant height in Miami and Las Vegas. The cooling load in Las Vegas and Miami (unlike Vancouver and Toronto) raised slightly by increasing the plant height when the plant coverage was low (LAI = 2). However, the results expressed that the cooling load decreased by increasing the plant height in all four cities when the LAI was high. Generally, the plant height was not as effective as LAI in the reduction of cooling load. By contrast, the heating load increased as the taller plants resulted in more shading on the soil surface. It is important to notice that the increase in heating load with plant height was slightly lower when the LAI was lower. This is justifiable as the surface of the green roof was exposed to more solar radiation when the plant coverage was less. As Las Vegas and Miami are cooling dominated climates, the annual energy savings with a 30 cm plant is more than with a 10 cm plant. However, in Toronto and Vancouver (which have heating dominated climates), the 10 cm plant had better performance on the building energy consumption. The energy performance of a conventional roof with and without insulation was also compared with green roofs. It is clear that by optimizing the green roof system the thermal performance of the green roof was considerably better than an uninsulated conventional roof. Additionally, the green roof had a better performance in the reduction of cooling load compared to the conventional roofs. It should be noted that the plant height and type of plant had an effect on the snow accumulation on green roofs as mentioned in previous sections. This is important as it has a major impact on the energy performance of green roofs and should be considered in the evaluation of green roofs especially for cold climates like Toronto that experience high snowfall during the winter.

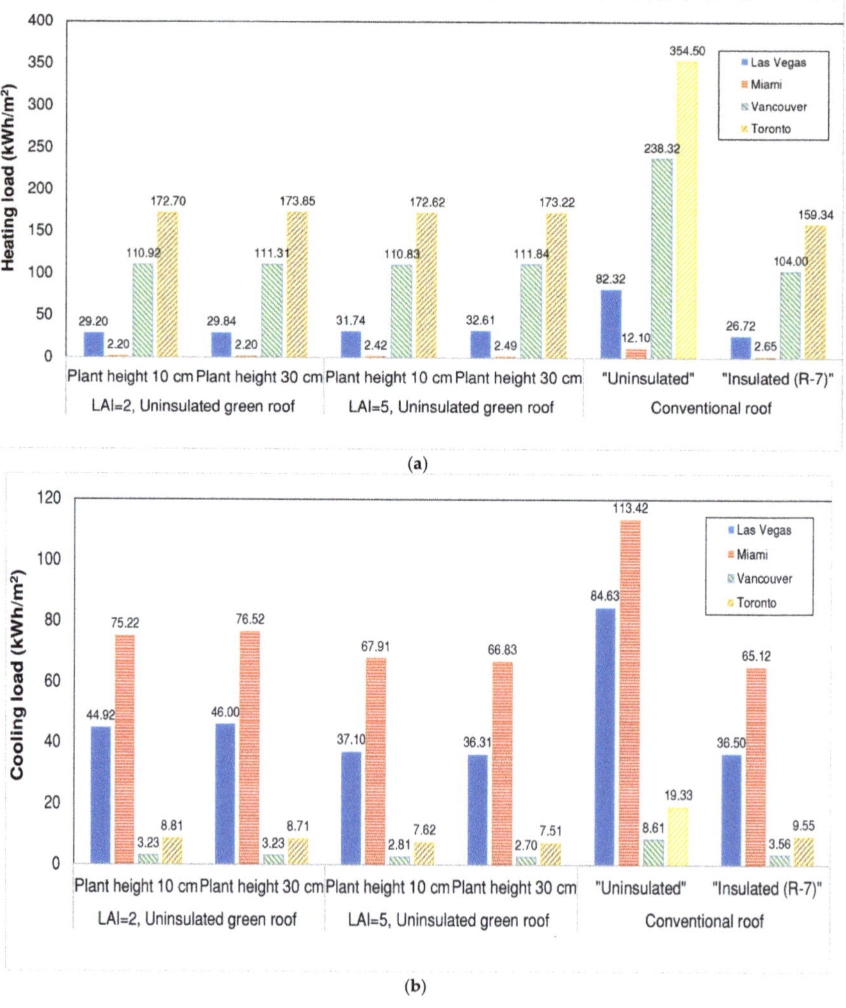

Figure 9. Sensitivity of (**a**) heating load and (**b**) cooling load, with LAI and plant height in different climates.

3.6. Influence of Thermal Insulation on an Optimized Green Roof

In this section an optimized green roof based on parametric studies performed in previous sections was considered. Figure 10 shows the role of thermal insulation on the thermal performance of the school building with optimized design parameters of a green roof as described above. The results demonstrated that thermal insulation would be more effective for the heating dominated climates. The heating load of an optimized vegetated roof in Miami and Las Vegas changed minimally with thermal insulation compared to Vancouver and Toronto. For example, in Miami the uninsulated green roof has almost equal heating load with insulated conventional roof. However, in Las Vegas the heating load for the insulated green roof is slightly better than an uninsulated green roof.

Nevertheless, the uninsulated vegetated roofs in Vancouver and Toronto have considerably higher heating loads than insulated vegetated roofs. It can be concluded that vegetated roofs cannot replace the thermal insulation in heated dominated climates like Canada. However, the insulated green roof with a 5 cm thermal insulation has almost the same heating load and lower cooling load compared to the conventional roof with a 10 cm thermal insulation and without a green roof system in these two cities. This means that a green roof could be an appropriate retrofit choice in Canada instead of adding extra insulation to the roof.

On the other hand, green roofs are more effective at reducing cooling loads and the uninsulated green roof has lower cooling loads than the insulated roofs in all cities except Miami. The cooling load decreased with insulation in Miami as the thermal storage was not as effective as it was in other cities because of the low oscillation of ambient air temperature in this warm-humid climate zone. As a result, adding thermal insulation could reduce the cooling load in this climate. In other cities, by increasing the thermal insulation from 0 to 100 mm the cooling load of the vegetated roof increased. Increasing the thermal insulation levels decouples the indoor environment from the outdoor environment and reduces the ability of the vegetated roof to remove heat from the building. Thus, it can be concluded that there are no benefits from continuously adding insulation to reduce the cooling load.

Table 9 presents the total energy saving of the optimized green roof in comparison with the conventional roof. Here, the conventional roof is considered the reference roof type. It is seen that the annual energy saving varies among the four selected cities. The annual energy consumption with an optimized green roof saved more in heating dominated cities than the cooling dominated cities. The uninsulated optimized green roof performed better than the insulated conventional roof in cooling dominated climates such as Miami and Las Vegas. For instance, installation of optimized green roof assembly on the uninsulated conventional roof saved almost 41% and 29% of annual energy consumption in Las Vegas and Miami, respectively.

Table 9. Comparing the energy consumption (kWh/m^2) of conventional roof (CR) and green roof (GR) at different level of thermal insulation.

City	Energy Consumption	Without Insulation			5 cm Insulation			10 cm Insulation		
		GR	CR	Diff %	GR	CR	Diff %	GR	CR	Diff %
Vancouver	Heating	110.83	238.32	53.49	91.11	104.00	12.39	87.14	92.52	5.81
	Cooling	2.81	8.61	67.36	2.90	3.56	18.54	2.97	3.26	8.89
	Total	227.84	382.73	40.46	207.33	221.07	6.22	203.01	209.00	2.86
Toronto	Heating	172.62	354.50	51.31	139.82	159.34	12.25	132.50	141.22	6.17
	Cooling	7.62	19.33	60.57	7.74	9.55	18.95	7.91	8.74	9.49
	Total	295.51	513.80	42.48	261.46	285.14	8.30	254.70	264.34	3.64
Las Vegas	Heating	32.61	82.32	60.38	26.72	29.44	9.23	25.54	26.66	4.20
	Cooling	36.31	84.63	57.09	36.50	43.82	16.70	36.72	40.74	9.86
	Total	183.10	309.93	40.92	177.94	190.70	6.69	177.10	183.73	3.61
Miami	Heating	2.49	12.10	79.42	2.23	2.65	15.85	2.22	2.40	7.50
	Cooling	66.83	113.42	41.07	65.12	71.94	9.48	64.82	68.30	5.09
	Total	185.34	260.11	28.74	183.50	192.22	4.54	183.14	187.44	2.29

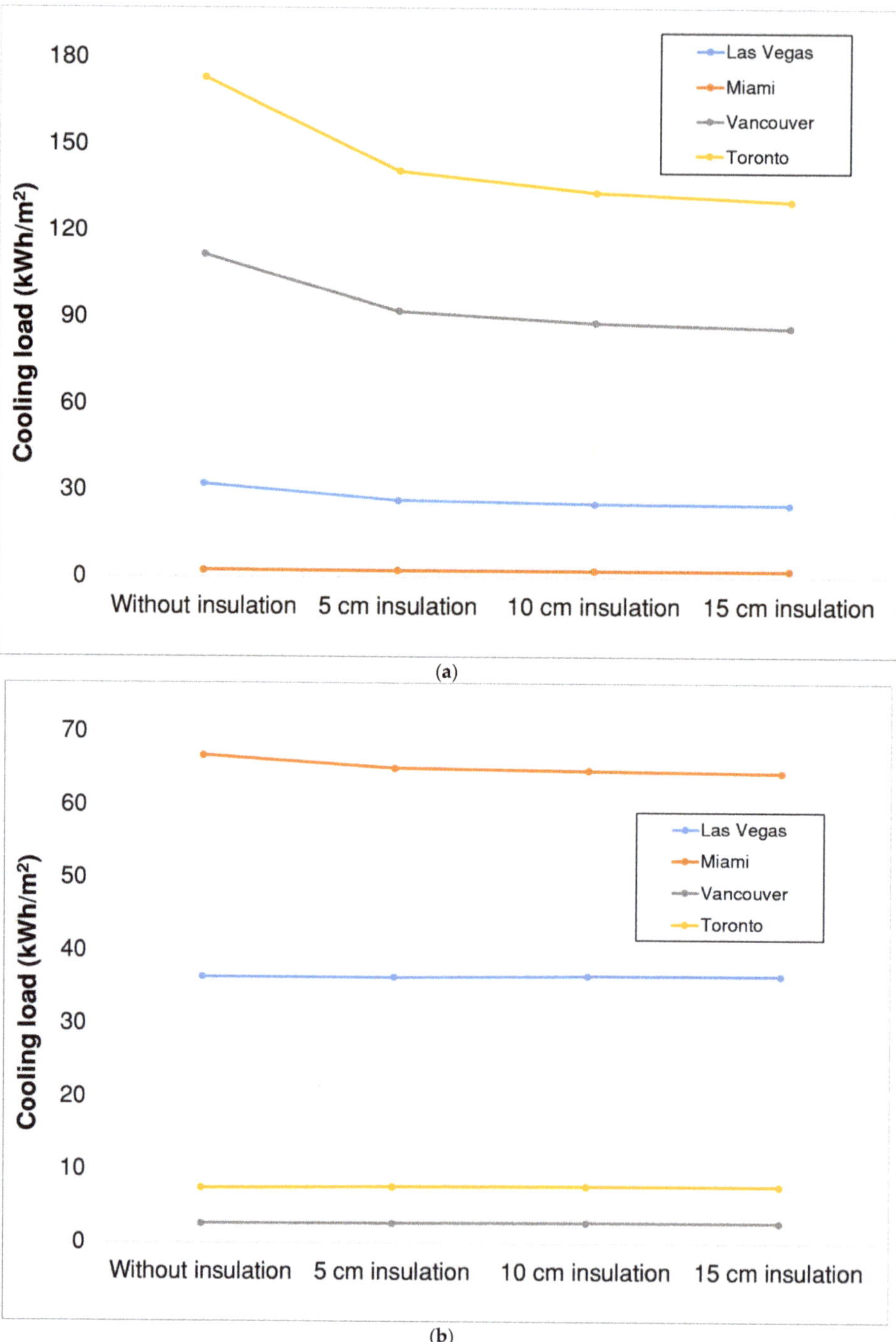

Figure 10. Sensitivity of (**a**) heating load and (**b**) cooling load, with thermal insulation for the optimized green roof in different climates.

Additionally, the optimized green roof without insulation saved energy almost 4% more than the poorly insulted conventional roof. Ultimately, the total energy consumption of the school building with the uninsulated green roof was slightly lower than the well-insulated conventional roof in both cities. It can be concluded that in Miami and Las Vegas, the optimized green roof is an appropriate replacement for the insulated conventional roofs.

However, in heating dominated cities, such as Vancouver and Toronto, an uninsulated optimized green roof is not a good choice for replacing the insulated conventional roof. It can be a suitable substitute for retrofitting the building with a poorly insulated roof. For instance, the heating load was reduced by 12.5% in a retrofitted optimized green roof compared to a conventional roof in both cities. Additionally, the annual energy consumption was saved by 8.30% and 6.22% in Toronto and Vancouver, respectively. Hence, in cold climates the thermal performance of the school building could be retrofitted by the green roof instead of adding extra thermal insulation to the conventional roof.

4. Conclusions

This study used energy modeling software tool EnergyPlus 8.8 to evaluate the influence of leaf area index (LAI), plant height, leaf albedo, substrate properties, and the roof's thermal insulation on the thermal performance of a secondary school building with extensive green roof in four cities: Las Vegas, NV (USA) and Miami, FL (USA), Toronto, ON (Canada) and Vancouver, BC (Canada). The first two cities are cooling dominated cities while the latter two are heating dominated cities. The following conclusions can be made from this study:

1. The effect of the substrate on the heating load is strongly related to its thermal conductivity. However, the substrate effect on the cooling loads depends on its thermal diffusivity.
2. For the uninsulated green roof, the light-weight substrate had considerably better performance than heavy-weight substrate in all four cities. However, for the insulated green roof there was a small difference between the energy performance of the green roof with a light-weight and heavy-weight substrate.
3. The leaf area index (LAI) is one of the most influential parameters and its effect on the energy performance of a green roof is climate specific. It has major impacts on the cooling reduction in all four selected cities. However, the effect of LAI on the heating load is dependent on the climatic condition. With regard to energy saving, the optimum LAI is 5 for the four selected cities.
4. Thicker soil has better energy savings in all four cities. In summer, the cooling effect of LAI reduces the effectiveness of soil thickness on the cooling load reduction. However, in the heating season, soil thickness has a large effect on the heating load reduction. The thicker soil (15 cm) has better thermal performance compared to the thinner soil (7.5 cm).
5. The thermal performance of plant height is affected by the LAI and climate zone of the region. Plant height should be optimal at 30 cm, except in Toronto and Vancouver, where it should be 10 cm.
6. The effect of seasonal variation of LAI on both cooling and heating loads should be considered.
7. Plant albedo has the least effect on the thermal performance of the school. For the heating dominated climate, darker leaves with an albedo of 0.11 and for the cooling dominated climates lighter leaves, with an albedo of 0.23 would be beneficial.
8. Seasonal variation of plant albedo has a small effect on cooling and heating load variation.
9. The green roofs cannot replace thermal insulation to reduce heating loads in cities that experience cold winters. However, it can retrofit the energy performance of the poorly insulated buildings in heating dominated climates.

In addition to the aforementioned conclusions, it has to be noted that that energy saving is one of the benefits of green roofs. Eventually, other environmental benefits of green roofs should also be considered to have a comprehensive evaluation for green roofs' performance in different climate zones.

Author Contributions: Conceptualization, P.M.; Formal analysis, M.M.; Funding acquisition, P.M.; Investigation, M.M.; Methodology, M.M. and P.M.; Project administration, P.M.; Validation, M.M.; Writing—review and editing, C.V.

Funding: This research received no external funding.

Acknowledgments: The authors would like thank the four reviewers for their insightful reviews that greatly improved this paper.

Conflicts of Interest: The authors declare no conflict of interest.

References

1. Vera, S.; Pinto, C.; Tabares-Velasco, P.C.; Bustamante, W.; Victorero, F.; Gironás, J.; Bonilla, C.A. Influence of vegetation, substrate, and thermal insulation of an extensive vegetated roof on the thermal performance of retail stores in semiarid and marine climates. *Energy Build.* **2017**, *146*, 312–321. [CrossRef]

2. Valipour, M. Future of agricultural water management in Africa. *Arch. Agron. Soil Sci.* **2015**, *61*, 907–927. [CrossRef]

3. Valipour, M. Land use policy and agricultural water management of the previous half of century in Africa. *Appl. Water Sci.* **2015**, *5*, 367–395. [CrossRef]

4. Jaffal, I.; Ouldboukhitine, S.-E.; Belarbi, R. A comprehensive study of the impact of green roofs on building energy performance. *Renew. Energy* **2012**, *43*, 157–164. [CrossRef]

5. Getter, K.L.; Rowe, D.B.; Andresen, J.A. Quantifying the effect of slope on extensive green roof stormwater retention. *Ecol. Eng.* **2007**, *31*, 225–231. [CrossRef]

6. Vijayaraghavan, K. Green roofs: A critical review on the role of components, benefits, limitations and trends. *Renew. Sustain. Energy Rev.* **2016**, *57*, 740–752. [CrossRef]

7. Teemusk, A.; Mander, Ü. Rainwater runoff quantity and quality performance from a greenroof: The effects of short-term events. *Ecol. Eng.* **2007**, *30*, 271–277. [CrossRef]

8. Berndtsson, J.C. Green roof performance towards management of runoff water quantity and quality: A review. *Ecol. Eng.* **2010**, *36*, 351–360. [CrossRef]

9. Alexandri, E.; Jones, P. Temperature decreases in an urban canyon due to green walls and green roofs in diverse climates. *Build. Environ.* **2008**, *43*, 480–493. [CrossRef]

10. Karachaliou, P.; Santamouris, M.; Pangalou, H. Experimental and numerical analysis of the energy performance of a large scale intensive green roof system installed on an office building in Athens. *Energy Build.* **2016**, *114*, 256–264. [CrossRef]

11. Takebayashi, H.; Moriyama, M. Surface heat budget on green roof and high reflection roof for mitigation of urban heat island. *Build. Environ.* **2007**, *42*, 2971–2979. [CrossRef]

12. van Renterghem, T.; Botteldooren, D. In-situ measurements of sound propagating over extensive green roofs. *Build. Environ.* **2011**, *46*, 729–738. [CrossRef]

13. Teemusk, A.; Mander, Ü. Greenroof potential to reduce temperature fluctuations of a roof membrane: A case study from Estonia. *Build. Environ.* **2009**, *44*, 643–650. [CrossRef]

14. Brenneisen, S. Space for urban wildlife: Designing green roofs as habitats in Switzerland. *Urban. Habitats* **2006**, *4*, 27–36.

15. Brenneisen, S. *The Benefits of Biodiversity from Green Roofs—Key Design Consequences*; University of Applied Sciences: Wadenswil, Switzerland, 2003; pp. 1–9.

16. Williams, N.S.; Lundholm, J.; MacIvor, J.S. Do green roofs help urban biodiversity conservation? *J. Appl. Ecol.* **2014**, *51*, 1643–1649. [CrossRef]

17. Ascione, F.; Bianco, N.; de'Rossi, F.; Turni, G.; Vanoli, G.P. Green roofs in European climates. Are effective solutions for the energy savings in air-conditioning? *Appl. Energy* **2013**, *104*, 845–859. [CrossRef]

18. Abuseif, M.; Gou, Z. A review of roofing methods: Construction features, heat reduction, payback period and climatic responsiveness. *Energies* **2018**, *11*, 3196. [CrossRef]

19. Getter, L.K.; Rowe, D.B.; Cregg, B.M. Solar radiation intensity influences extensive green roof plant communities. *Urban. For. Urban. Green.* **2009**, *8*, 269–281. [CrossRef]

20. Heusinger, J.; Stephan, W. Extensive green roof CO^2 exchange and its seasonal variation quantified by eddy covariance measurements. *Sci. Total Environ.* **2017**, *607*, 623–632. [CrossRef]

21.	Akther, M.; He, J.; Chu, A.; Huang, J.; Van Duin, B. A review of green roof applications for managing urban stormwater in different climatic zones. *Sustainability* **2018**, *10*, 2864. [CrossRef]

22.	Foustalieraki, M.; Assimakopoulos, M.; Santamouris, M.; Pangalou, H. Energy performance of a medium scale green roof system installed on a commercial building using numerical and experimental data recorded during the cold period of the year. *Energy Build.* **2017**, *135*, 33–38. [CrossRef]

23.	DeNardo, J.C.; Jarrett, A.; Manbeck, H.; Beattie, D.; Berghage, R. Green roof mitigation of stormwater and energy usage. In Proceedings of the ASAE Annual Meeting, American Society of Agricultural and Biological Engineers, Las Vegas, NV, USA, 27–30 July 2003; p. 1.

24.	Bevilacqua, P.; Mazzeo, D.; Bruno, R.; Arcuri, N. Experimental investigation of the thermal performances of an extensive green roof in the Mediterranean area. *Energy Build.* **2016**, *122*, 63–79. [CrossRef]

25.	Babak, R.; Tenpierik, M.J.; van den Dobbelsteen, A. The impact of greening systems on building energy performance: A literature review. *Renew. Sustain. Energy Rev.* **2015**, *45*, 610–623.

26.	Wong, N.H.; Cheong, D.K.W.; Yan, H.; Soh, J.; Ong, C.; Sia, A. The effects of rooftop garden on energy consumption of a commercial building in Singapore. *Energy Build.* **2003**, *35*, 353–364. [CrossRef]

27.	Theodosiou, T.G. Summer period analysis of the performance of a planted roof as a passive cooling technique. *Energy Build.* **2003**, *35*, 909–917. [CrossRef]

28.	Vera, S.; Pinto, C.; Victorero, F.; Bustamante, W.; Bonilla, C.; Gironás, J.; Rojas, V. Influence of plant and substrate characteristics of vegetated roofs on a supermarket energy performance located in a semiarid climate. *Energy Procedia* **2015**, *78*, 1171–1176. [CrossRef]

29.	Sailor, D.J.; Elley, T.B.; Gibson, M. Exploring the building energy impacts of green roof design decisions—A modeling study of buildings in four distinct climates. *J. Build. Phys.* **2012**, *35*, 372–391. [CrossRef]

30.	Getter, K.L.; Rowe, D.B.; Andresen, J.A.; Wichman, I.S. Seasonal heat flux properties of an extensive green roof in a Midwestern US climate. *Energy Build.* **2011**, *43*, 3548–3557. [CrossRef]

31.	Silva, C.M.; Gomes, M.G.; Silva, M. Green roofs energy performance in Mediterranean climate. *Energy Build.* **2016**, *116*, 318–325. [CrossRef]

32.	Mukherjee, S.; la Roche, P.; Konis, K.; Choi, J.H. Thermal performance of green roofs: A parametric study through energy modeling in different climates. In Proceedings of the Passive Low Energy Architecture Conference PLEA, Munich, Germany, 10–12 September 2013.

33.	Gomes, M.G.; Silva, C.M.; Valadas, A.S.; Silva, M. Impact of Vegetation, Substrate, and Irrigation on the Energy Performance of Green Roofs in a Mediterranean Climate. *Water* **2019**, *11*, 2016. [CrossRef]

34.	Lundholm, J.T.; Weddle, B.M.; MacIvor, J.S. Snow depth and vegetation type affect green roof thermal performance in winter. *Energy Build.* **2014**, *84*, 299–307. [CrossRef]

35.	Tang, X.; Qu, M. Phase change and thermal performance analysis for green roofs in cold climates. *Energy Build.* **2016**, *121*, 165–175. [CrossRef]

36.	Akbari, H.; Konopacki, S. Energy effects of heat-island reduction strategies in Toronto, Canada. *Energy* **2004**, *29*, 191–210. [CrossRef]

37.	Liu, K.; Baskaran, B. Thermal performance of green roofs through field evaluation. In Proceedings of the First North American Green Roof Infrastructure Conference, Awards and Trade Show, Chicago, IL, USA, 29–30 May 2003; pp. 29–30.

38.	Zhao, M.; Srebric, J. Assessment of green roof performance for sustainable buildings under winter weather conditions. *J. Cent. South. Univ.* **2012**, *19*, 639–644. [CrossRef]

39.	Connelly, M.; Liu, K. Green roof research in British Columbia: An overview. In Proceedings of the 3rd North American Green Roof Conference: Greening rooftops for sustainable communities, Washington, DC, USA, 4–6 May 2005; pp. 4–6.

40.	Berardi, U. The outdoor microclimate benefits and energy saving resulting from green roofs retrofits. *Energy Build.* **2016**, *121*, 217–229. [CrossRef]

41.	Niachou, A.; Papakonstantinou, K.; Santamouris, M.; Tsangrassoulis, A.; Mihalakakou, G. Analysis of the green roof thermal properties and investigation of its energy performance. *Energy Build.* **2001**, *33*, 719–729. [CrossRef]

42.	Zhao, M.; Tabares-Velasco, P.C.; Srebric, J.; Komarneni, S.; Berghage, R. Effects of plant and substrate selection on thermal performance of green roofs during the summer. *Build. Environ.* **2014**, *78*, 199–211. [CrossRef]

43.	la Roche, P.; Berardi, U. Comfort and energy savings with active green roofs. *Energy Build.* **2014**, *82*, 492–504. [CrossRef]

44. Managing energy costs in schools, A guide to energy conservation and savings for K-12 schools, Xcel Energy In. 2007. Available online: https://www.xcelenergy.com/staticfiles/xe/Marketing/Managing-Energy-Costs-Schools.pdf (accessed on 12 December 2019).
45. *Energy Efficiency in K-12 Schools & State Applications*; US Department of Energy: Washington, DC, USA, 2013.
46. Lemire, N. School and more. *Ashrae J.* **2010**, *52*, 34.
47. Ouf, M.M.; Mohamed, H.I. Energy consumption analysis of school buildings in Manitoba, Canada. *Int. J. Sustain. Built Environ.* **2017**, *6*, 359–371. [CrossRef]
48. Ouf, M. The effect of occupancy on school buildings' energy consumption in Manitoba. Ph.D. Dissertation, University of Manitoba, Winnipeg, MB, Canada, 2016.
49. U.S. DOE. Energyplus engineering reference. In *The Reference to Energyplus Calculations*; U.S. DOE: Washington, DC, USA; p. 2010.
50. Sailor, D.J. A green roof model for building energy simulation programs. *Energy Build.* **2008**, *40*, 1466–1478. [CrossRef]
51. Deru, M.; Field, K.; Studer, D.; Benne, K.; Griffith, B.; Torcellini, P.; Liu, B.; Halverson, M.; Winiarski, D.; Rosenberg, M.; et al. *US Department of Energy Commercial Reference Building Models of the National Building Stock*; National Renewable Energy Laboratory: Jefferson County, CO, USA, 2011; pp. 1–118.
52. U.S. DOE. Commercial Prototype Building Models, Department of Energy, United States. Available online: https://www.energycodes.gov/development/commercial/prototypemodels (accessed on 24 October 2019).
53. EnergyPlus. Energy Simulation Software. 2013. Available online: /https://www.energy.gov/eere/buildings/energyplus (accessed on 23 September 2019).
54. Wilcox, S.; Marion, W. *User's Manual for TMY3 Data Sets*; National Renewable Energy Laboratory: Golden, CO, USA, 2008.
55. Logics, N. *Canadian Weather for Energy Calculations*; Users Manual and CD-ROM; Environment Canada: Downsview, ON, Canada, 1999.
56. Sailor, D.J.; Bass, B. Development and Features of the Green Roof Energy Calculator (GREC). *J. Living Archit.* **2014**, *1*, 36–58.
57. Feng, C.; Meng, Q.; Zhang, Y. Theoretical and experimental analysis of the energy balance of extensive green roofs. *Energy Build.* **2010**, *42*, 959–965. [CrossRef]
58. Gagliano, A.; Detommaso, M.; Nocera, F. Assessment of the Green Roofs Thermal Dynamic Behavior for Increasing the Building Energy Efficiencies. In *Smart Energy Control Systems for Sustainable Buildings*; Springer: Cham, Switzerland, 2017; pp. 37–59.
59. Nektarios, P.A.; Ntoulas, N.; Nydrioti, E.; Kokkinou, I.; Bali, E.M.; Amountzias, I. Drought stress response of Sedum sediforme grown in extensive green roof systems with different substrate types and depths. *Sci. Hortic.* **2015**, *181*, 52–61. [CrossRef]
60. Schaap, M.G.; van Genuchten, M.T. A modified Mualem–van Genuchten formulation for improved description of the hydraulic conductivity near saturation. *Vadose Zone J.* **2006**, *5*, 27–34. [CrossRef]
61. Mahmoodzadeh, M. Impact of Extensive Green Roofs on Energy Performance of School Buildings in four North American Climates. Masc Dissertation, University of Victoria, Victoria, BC, Canada, 2018.
62. Sailor, D.; Hutchinson, D.; Bokovoy, L. Thermal property measurements for ecoroof soils common in the western US. *Energy Build.* **2008**, *40*, 1246–1251. [CrossRef]
63. Green Roof Model (EcoRoof). Available online: https://bigladdersoftware.com/epx/docs/8-3/engineering-reference/green-roof-model-ecoroof.html (accessed on 12 December 2019).
64. Yaghoobian, N.; Srebric, J. Influence of plant coverage on the total green roof energy balance and building energy consumption. *Energy Build.* **2015**, *103*, 1–3. [CrossRef]
65. Zeng, C.; Bai, X.; Sun, L.; Zhang, Y.; Yuan, Y. Optimal parameters of green roofs in representative cities of four climate zones in China: A simulation study. *Energy Build.* **2017**, *150*, 118–131. [CrossRef]
66. Boafo, F.E.; Kim, J.-T.; Kim, J.-H. Evaluating the impact of green roof evapotranspiration on annual building energy performance. *Int. J. Green Energy* **2017**, *14*, 479–489. [CrossRef]

Article

An Integrated Hydraulic and Hydrologic Modeling Approach for Roadside Bio-Retention Facilities

James Li *, Seyed Alinaghian, Darko Joksimovic and Lianghao Chen

Department of Civil Engineering, Ryerson University, Toronto, ON M5B 2K3, Canada;
Alex.Alinaghian@ryerson.ca (S.A.); darkoj@ryerson.ca (D.J.); lianghao.chen@peelregion.ca (L.C.)
* Correspondence: jyli@ryerson.ca; Tel.: +001-416-979-5345

Received: 20 March 2020; Accepted: 18 April 2020; Published: 27 April 2020

Abstract: Roadside bio-retention (RBR) facilities are low impact development practices, which control urban runoff primarily from road pavements. Using hydrologic models, such as the US EPA Storm Water Management Model (SWMM), RBR are typically designed with some fundamental assumptions, including where runoff completely enters the facilities and fully utilizes the whole surface area for percolation, detention, filtration, and infiltration to the surrounding soils. This paper highlights the importance of inlet hydraulics and the spatial distribution of inflow along a RBR, and proposes an integrated hydraulic and hydrologic modelling approach to simulate its overall runoff control performance. The integrated hydraulic/hydrologic modelling approach consists of three components: (1) A dual drainage hydrologic model to simulate runoff generation, runoff hydrographs entering and bypassing a storm inlet, and the outflow hydrograph from a fully utilized RBR; (2) a computational fluid dynamic model to determine the inflow distribution along a RBR; and (3) an overall runoff control performance analysis of RBR by considering the inlet efficiency, and the partially and fully utilized RBR during a storm event. A case study of an underground RBR in the City of Toronto was used to demonstrate the integrated modelling approach. It is concluded that; (1) inlet efficiency of a RBR will determine the overall runoff control performance; and (2) the inflow distribution will dictate the effective length of a RBR, which may affect the overall runoff control performance.

Keywords: bio-retention; green infrastructure; runoff control performance; storm inlet hydraulics; flow distribution hydraulics

---

1. Introduction

In order to control runoff problems (e.g., increased flooding, deterioration of water quality, receiving channel erosion) associated with urban development, low impact development (LID) aims at restoration of pre-development hydrology by utilizing spatially distributed LID practices throughout a drainage catchment [1]. LID practices have been recognized to harness natural hydrologic processes, in order to "manage rainfall at the source, using uniformly distributed decentralized micro-scale control"; [2] and can be divided into lot-based, road right-of-way-based, and area-based types [1]. Lot-based LID practices are facilities on a single parcel of land, such as a building lot. Right-of-way (ROW)-based LID practices are facilities within the area of land that is acquired for or devoted to the provision of a road [3]. Area-based LID practices, also known as multi-lot LIDs, are located on large pieces of land such as parks. Among all these different types of LIDs, ROW-based LIDs are the most difficult facilities to implement, due to lack of space and the layout of the road drainage and surrounding utility system. Nevertheless, storm runoff along ROW should be controlled because the total road drainage area of some highly urbanized cities or regions can be as large as adjacent building lots and open space. Several innovative structural techniques to control ROW runoff include bio-retentions, porous pavement, catch basin filters, sewer trench exfiltration systems, perforated sewers, and exfiltration tanks [1,4–6]. Among all these techniques, ROW-based LIDs, such as roadside bio-retentions (RBR) are becoming popular and can be integrated into roadside landscaping green space. RBR has two common types:

(1)   Conventional surface RBR with a surface plant soil layer

Storm runoff, generates from a section of the ROW, is captured by one or multiple inlets (e.g., curb cuts or openings) to a RBR, either on a sidewalk or the median of the road. It can be integrated into the neighborhood green space or flower beds. A typical RBR consists of the top vegetation layer, intermediate filter media (e.g., sandy loam), intermediate transition layer (e.g., coarse sand), a drainage layer (e.g., coarse sand and gravel) with, and without, a perforated underdrain collection pipe [7]. Runoff entering the RBR can temporarily pond at the surface layer and subsequently infiltrate into the various layers of the bio-retention cell. There may be an overflow pipe at the surface layer for excess runoff conveyance. The percolated runoff can go, either to the surrounding native soil, or an underdrain pipe connecting back to the road storm sewer [8].

(2)   Underground RBR without a surface layer (Figure 1)

The layout of this RBR is similar to the conventional one except the absence of the top soil-plant layer. This type of RBR is commonly used in highly urbanized area. The inlets can be in the form of a grate and/or curb opening catch basin, which conveys the captured runoff to the bio-retention cell through an upper perforated flow distribution pipe. The percolated runoff can go either to the surrounding native soil and/or an underdrain pipe connecting back to the road storm sewer. It is useful for highly urbanized areas, since it can be constructed under the sidewalk and part of the road without affecting parking space [9].

Current research studies on RBR, as summarized below, have focused primarily on runoff volume-controlled, water pollutant removal, enhanced soil mix, and clogging using laboratory, and field scale investigations and mathematical modeling.

(1)   Laboratory scale research

Laboratory studies of RBR using bio-retention boxes and columns have focused on hydrologic performance [10,11], water quality treatment [12], heavy metal removal efficiency [13,14], organic carbon removal efficiency [15,16], nutrient removal efficiency [17,18], effect of salts on performance [13,19,20], and root effect on infiltration rates [11].

(2)   Field scale research

Field studies of RBR have been conducted to investigate the effect of plant and fungal contributions to performance [21], soil phosphorus release [22], treatment efficiency of runoff highway bridge deck [23], heavy metal removal efficiency [24–26], nitrogen removal efficiency [27,28], greenhouse gas release [29,30], hydraulic properties and infiltration rates [31] and operation and maintenance [32,33].

(3)   Mathematical modeling research

Mathematical modeling studies of RBR have focused on sizing and optimization methods [25,34–36]; hydrologic performance [37–39], and hydraulic and hydrologic performance [40–44].

Research of storm inlet hydraulics has dated back to 1950s [45–47], continued through 1990s [48], and in the last 10 years [49–54]. However, the effect of inlet hydraulics on the overall performance of RBR has only been investigated recently [55,56]. This paper focuses on this research gap and further investigates the captured inflow distribution, along an underground RBR using laboratory experiments, computational fluid dynamic (CFD) simulations, and hydrological modelling. A case study of an underground RBR in Toronto (Figures 1 and 2) is used as a demonstration of the integrated hydraulic and hydrologic approach to model the overall performance evaluation of the RBR. The case study RBR was designed as a treatment LID without infiltration and all percolated runoff was collected by an underdrain to the street storm sewer.

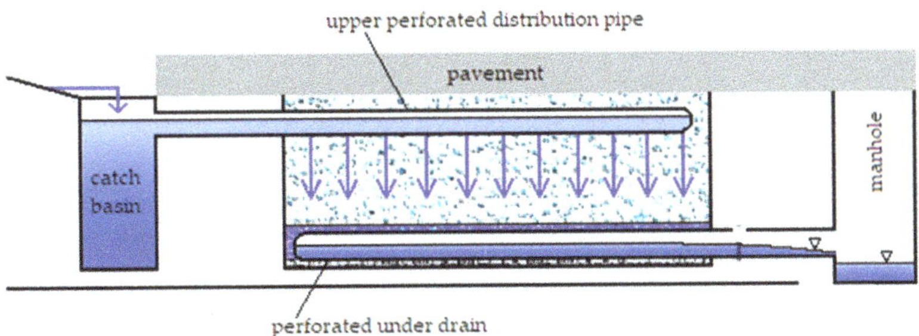

Figure 1. An underground RBR without a surface layer.

Figure 2. Construction phases of the Toronto underground RBR without a surface layer.

2. Materials and Methods

The integrated hydraulic and hydrologic modelling approach for the overall performance analysis of an underground RBR is based on the tracking of runoff from a road drainage area, which enters a storm inlet and the RBR, and exits the RBR by, either infiltration to the surrounding soil, or conveyance through a bottom underdrain to the storm sewer. It consists of three consecutive components: (1) Development of a dual drainage hydrologic model, which simulates runoff hydrographs entering and bypassing a RBR and the runoff control performance of a fully utilized RBR; (2) hydraulic simulations of the inflow distribution along a RBR; and (3) determination of the overall runoff control performance of a RBR with outflow adjustments for partially utilization of the RBR during a storm event. A case study of an underground RBR without a surface layer implemented at the City of Toronto is used to illustrate the integrated modelling approach.

2.1. A Dual Drainage Hydrologic Model for a Fully Utilized Underground RBR

SWMM [57] is a popular hydrological analysis model for urban drainage planning and design. The PCSWMM software (Computational Hydraulics International, Guelph, ON, Canada) [58], utilizing the SWMM computational engine, can be used to develop a dual drainage hydrologic model for an underground RBR. PCSWMM simulates runoff along a street and to a storm sewer by specifying the section characteristic of a street (e.g., length, width, curb height, cross-slope, bank-slope and bank-height) and the sewer characteristics (e.g., length, roughness, inlet and outlet elevations, cross-section and geometry). Runoff enters a storm inlet longitudinally along the road gutter (i.e., frontal flow) and laterally from the side of a road (i.e., side flow). The captured runoff by a storm inlet becomes the inflow to a RBR. The total capacity of storm inlets is normally designed to match the storm sewer capacity (e.g., runoff rate associated with a storm of two or five year return period in Canada). For each type of inlets, there is an inlet capacity curve (e.g., Figure 3), which defines the total intercepting flow (e.g., outflow in Figure 3 on the Y-axis) at different hydraulic heads above the inlet (e.g., X-axis in Figure 3). Using an inlet capacity curve, a dual drainage model can be developed to determine the captured and bypassed runoff hydrographs at a storm inlet. The PCSWMM also has a LID module which can be used to model the infiltration and the underdrain flow of a RBR. As a result, PCWMM has the flow simulation mechanisms to track runoff generated from a road drainage area, the bypassed and captured runoff at a storm inlet, and the inflow and outflow of an underground RBR.

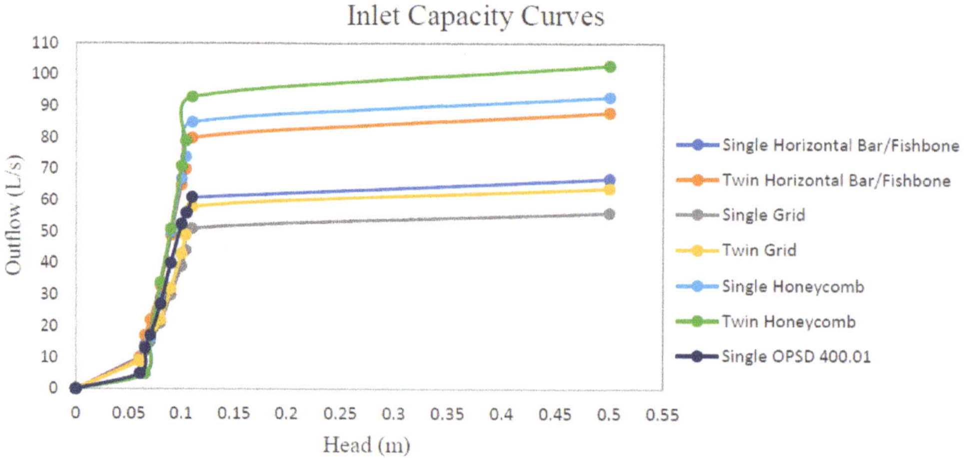

Figure 3. Toronto inlet capacity curves for different catch basins.

As an example, the dual drainage model of the Toronto RBR is shown in Figure 4 and Tables 1–3. Chicago design storms (Figure 5) with different return periods (2-, 5-, 10-, 25-, and 100-year) were used as rainfall inputs to the PCSWMM dual drainage model, which simulated the runoff hydrograph from the drainage area, the storm inlet bypass hydrograph, the inflow and outflow hydrographs of the RBR for different design storm conditions. All these simulated hydrographs were used to determine the overall runoff control performance of the RBR.

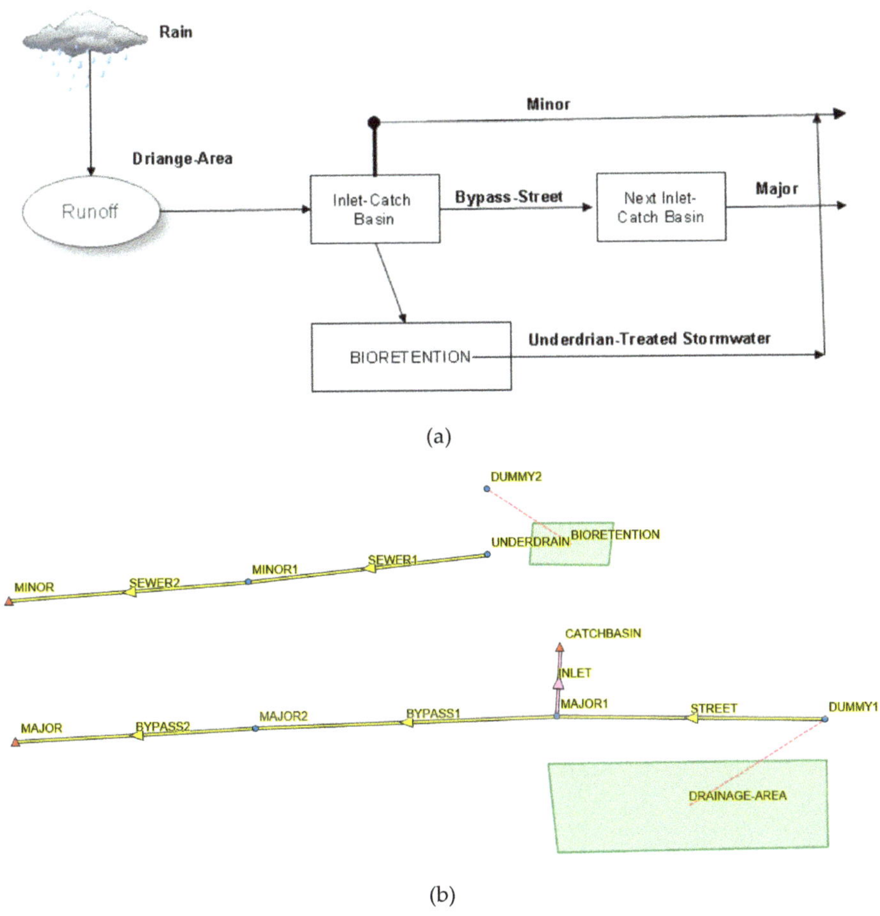

Figure 4. (**a**) Flow pathways through the RBR; (**b**) The dual drainage model of the Toronto RBR.

Table 1. Input parameters for the drainage area.

Inputs	Drainage-Area	Bioretention
Area (ha)	0.1074	0.0059
Flow Length (m)	74.554	18.081
Width (m)	14.352	3.2
Slope (%)	0.8	0.5
Imperv. (%)	100	0
N Imperv	0.011	0.011
N Perv	0.1	0.24
Dstore Imperv (mm)	1.25	0.127
Dstore Perv (mm)	2.5	0.508
Zero Imperv (%)	100	0

Table 2. Input parameters for the street and the sewer.

Inputs	Street	Sewer1
Length (m)	74.552	70
Roughness	0.016	0.011
Inlet Elevation (m)	91.03	90.43
Outlet Elevation (m)	90.43	89.87
Cross-Section	Irregular	Circular
Geom 1 (m)	Transect: Street1	0.5

Table 3. Input parameters for the bioretention.

Surface Layer			
Berm Height (mm)	Vegetation Volume (fraction)	Surface Roughness (Manning's n)	Surface Slope (%)
0.001	0	0.1	0.001

Soil Layer						
Thickness (mm)	Porosity (volume fraction)	Field Capacity (volume fraction)	Witling Point (volume fraction)	Conductivity (mm/h)	Conductivity Slope	Suction Head (mm)
600	0.437	0.105	0.047	29.97	10	60.96

Storage Layer			
Thickness (mm)	Void Ratio (voids/solids)	Seepage Rate (mm/h)	Clogging Factor
600	0.5	0	0

Underdrain		
Drain Coefficient (mm/h)	Drain Exponent	Drain Offset Height (mm)
339.17	0.5	0

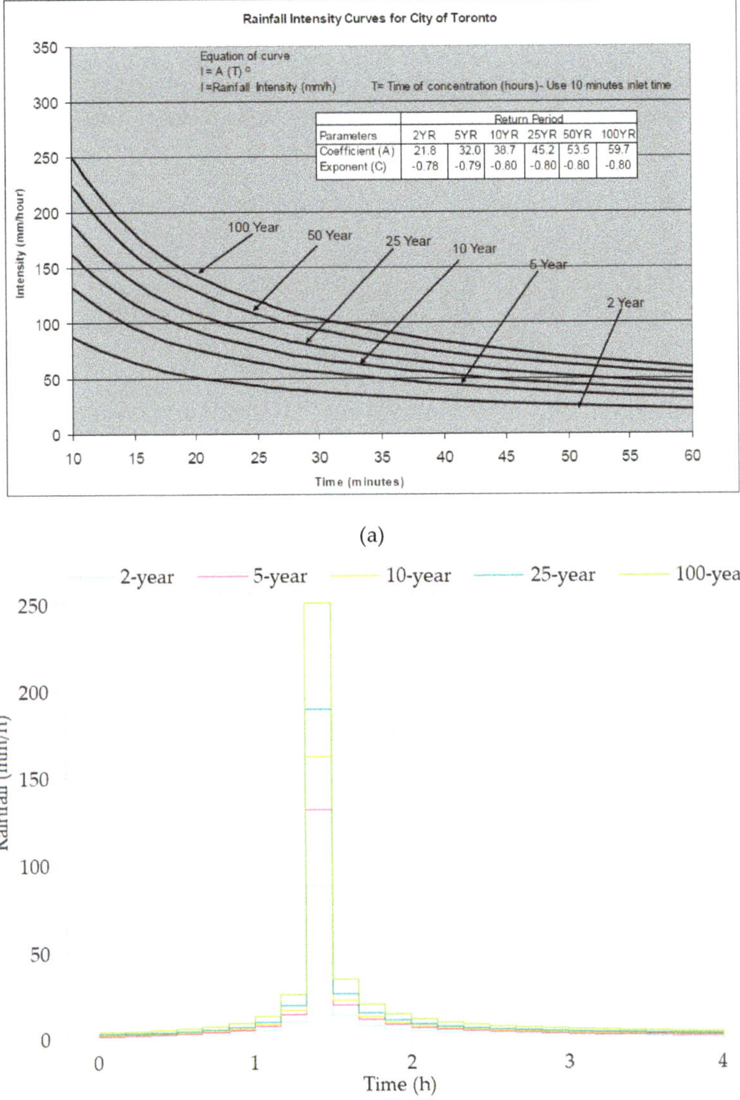

(a)

(b)

Figure 5. (**a**) Toronto Intensity-Duration-Frequency curves; and (**b**) Chicago Design Storms.

2.2. Hydraulic Simulations of the Inflow Distribution along an Underground RBR

As shown in Figure 1, the captured runoff from a storm inlet enters the RBR through an upper perforated pipe. Depending on the perforation size, its orientation around the circumstance, the distance between perforations, and the pipe slope, the runoff can distribute non-uniformly along the pipe. The water profiles under different inflow rates determine the spread of runoff along a RBR. If the water profile ends before the end of the pipe, due to the perforation arrangement, only a partial portion of the RBR is utilized for detention, filtration, and chemical/biological exchange of runoff quantity and quality. In order to maximize the usage of a RBR for runoff management, the perforation arrangement should be designed properly to spread the flow throughout a RBR under a design storm condition. Nevertheless, it is seldom considered in the design of an RBR.

The perforated flow distribution pipe is an important feature of LID, including the underground RBR. Duchene and McBean [59] investigated the full pipe discharged characteristics of smooth-wall and corrugated perforated pipes for use in infiltration trenches, and calibrated the orifice coefficient (0.65) of an equivalent orifice with good agreement with experimental measurements. Afrin et al. [60] developed CFD models using the software FLUENT (ANSYS Inc., Canonsburg, PA, USA) to simulate the hydraulic performance of perforated pipe underdrains surrounded by loose aggregates. Liu et al. [61] developed analytical equations for outflow, along the flow in a perforated fluid distribution pipe, based on momentum conservation method. Qin et al. [62] studied the flow velocity distribution along perforated pipes and found the laying slope could be used to promote uniform velocity distribution along the perforated pipes. Paul et al. [63] applied the EPANET software (US Environmental Protection Agency Research, Durham, NC, USA) and assessed discharge uniformity from perforated pipes in constructed wetlands.

All of the above research studies concentrated on full pipe discharge characteristics. On the other hand, this research focuses on the partial flow characteristics of a perforated distribution pipe, resulting in under-utilization of a RBR. A physical hydraulic model was used to collect data for the calibration of a CFD water profile model, which was subsequently used to simulate the flow profiles along an RBR under different inflow conditions.

2.2.1. Experimental Investigations of Water Profiles along a Perforated Pipe

A physical hydraulic model was used to analyze the water profiles along a 6.7 m long section of a 150 mm corrugated High-Density Polyethylene (HDPE) perforated pipe supplied by Armtec Inc. as per American Association of State Highway and Transportation Officials (AASHTO) M252 Type C standard. Four perforations were cut into the groove of the pipe, at 90 degrees apart, at the factory. Two perforation sizes were investigated: 25 mm by 25 mm, and 50 mm by 50 mm. The total number of perforation sets were 275. Additional, the pipe was wrapped in woven geotextile (important for application in the soil layer of a RBR). As shown in Figure 6, the experimental apparatus comprised of an upper reservoir, made of a 30 cm semi-transparent barrel with a 150 mm rigid pipe protruding out near the bottom to serve as an outlet. The lower reservoir was made up of four interconnecting plastic water tanks of 1.2 m in diameter. The perforated pipe was connected to the upstream upper reservoir using a watertight rubber corrugated rigid pipe coupling (Part No.: 1070-66), supplied by Fernco Connectors Ltd. (Sarnia, ON, Canada), while its downstream end was capped using the same coupling. Four piers supported the whole setup, while the perforated pipe was fastened to two by four wood studs by plastic straps. The perforated pipe slope was adjusted to almost level. Water circulation was provided by an IntelliflowXF variable speed pump (Pentair Inc., Minneapolis, MN, USA) through a 50 mm Poly Vinyl Chloride (PVC) rigid pipe. The flow rate was measured downstream of the pump using a F-1000 paddle wheel flowmeter manufactured by the Blue-White industries, Ltd. (Huntington Beach, CA, USA). As the water level at the upper reservoir rose, the flow inside the perforated pipe exfiltrated through the perforations, collected by a gutter system and returned to the lower reservoir. Four 6 mm holes were drilled manually at the bottom of the pipe at 1.5 m apart, starting from the entrance of the perforated pipe. The screw thread at the top of Omega PX309 pressure transducers (Omega Environmental, St-Eustache, QC, Canada) screwed tightly into the four 6 mm holes and wired to a Sutron 9210 Xlite Data Logger (OTT Hydromet GmbH, Kempten, Germany). Figure 7 shows a photograph of the experimental setup. Repeated measurements of the water depths (at least 3 times) were taken at 5 min intervals up to 20 min along the perforated pipe for flows ranging from 0.63 to 10.28 L/s. The measured water profiles under different inflows were used to calibrate a water profile simulation model described in the next subsection.

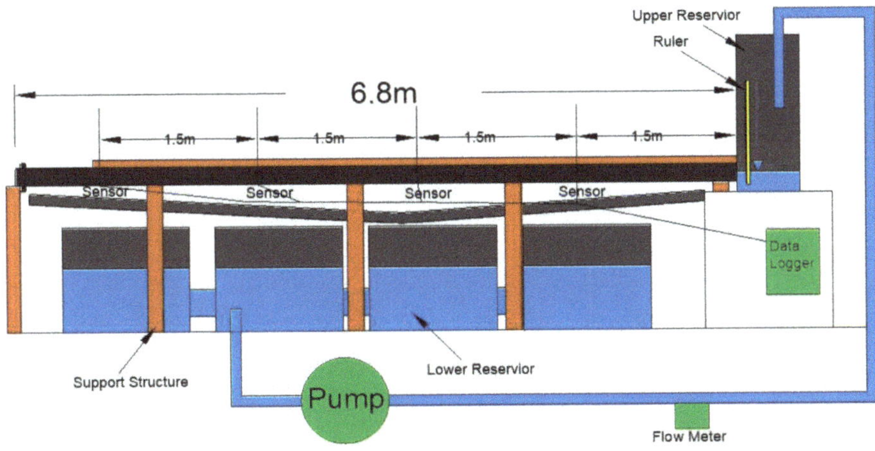

Figure 6. Schematic layout of the exfiltration pipe experimental apparatus.

Figure 7. Photograph of the experimental apparatus.

2.2.2. CFD Simulation of Water Profiles Long a Perforated Pipe

The FLOW-3D software [64] (Flow Science Inc., Santa Fe, NM, USA) was used to model the water profiles along the experimental perforated pipe under five inflow rates (from 0.62 to 10.28 L/s). Based on the finite-volume method, the Reynolds-averaged Navier Stokes equation was solved by the FLOW-3D. The computational domain of the pipe was divided into rectangular three dimensional computational grid of cells. The geometry of the perforated pipe and the perforations were drawn by AutoCAD-3D and imported to the FLOW-3D software. A large block of mesh was applied for the entire perforated pipe and small meshes were used for the perforation area. Since two-phase simulations were not used, the boundary condition in this research was the inner wall of the perforated pipe. The FLOW-3D model was calibrated using the measured water profiles in the experiments. The calibrated FLOW-3D model was then used to simulate the lengths of the partially filled perforated pipe of the Toronto RBR under different inflow rates. These simulated lengths were then used to determine the outflow adjustments for the Toronto RBR under partial inflow distribution conditions.

2.3. Overall Runoff Control Performance Analysis of a RBR with Outflow Adjustment for the Partially Filled Distribution Pipe

Many hydrologic models, which were developed to simulate the runoff control performance of RBR, were based on the assumption of a fully utilized RBR [27–29,40–44]. Depending upon the inflow rate, the inflow may, or may not, reach the full length of a RBR. For the underground RBR in Figure 1, it is important to determine the critical inflow rate at which the water profile will extend to the full length of a RBR, resulting in the full utilization of a RBR for runoff control. Any inflow less than this critical inflow rate will affect the runoff control performance because only a fraction of the RBR will be utilized for runoff control.

The calibrated CFD model, described in the previous subsection, was used to determine the critical inflow rate of the Toronto RBR at which the inflow would spread over the entire length of the perforated distribution pipe. Using the dual drainage model described in Section 2.1, the outflow hydrographs of different effective lengths of the Toronto RBR were analyzed. If the inflow rate of the RBR was less the critical inflow rate, the outflow hydrograph of a partially utilized RBR was used. Otherwise, the outflow hydrograph of the fully utilized RBR was adopted. Finally, the overall runoff control performance (e.g., volume and peak flow controlled) for different design storm conditions was determined by comparing the total runoff and the adjusted outflow hydrographs of the RBR.

3. Results

For the Toronto RBR case study, the integrated hydraulic and hydrologic approach was used to investigate: (1) Effect of catch basin types on the bypassed and captured runoff hydrographs and the runoff control performance; and (2) effect of inflow distribution on the overall runoff control performance.

3.1. Effect of Catch Basin Types on the Bypassed and Captured Runoff Hydrographs

Four types of inlets [65] were investigated for the 18 m long RBR: (a) single horizontal bar/fishbone catch basins; (b) OPSD400.01 catch basin; (c) single and twin grid catch basins; and (d) single and twin honeycomb catch basins. The PCSWMM model developed for the Toronto RBR was used to analyze (1) the bypassed and captured hydrographs at the catch basin; (2) the inlet and outlet hydrographs of the RBR; and (3) the peak runoff rate reduction (i.e., % reduction of the peak runoff rate) and the runoff volume controlled (% reduction of the total runoff volume) of the RBR for different design storms.

3.1.1. Single and Twin Horizontal Bar/Fishbone Catch Basins

Figure 8 shows the inlet and outlet hydrographs of the RBR and the bypassed hydrograph at the catch basin for 4-h design storms of 2 to 100 year return periods. The bypassed hydrographs (Figure 8b) of 2-, and 100-year storms have peak flow rates 2 and 10 times higher than those of the inlet hydrographs (Figure 8a). The peak runoff, captured by the catch basin, occurs after almost 90 min for all storms, while the peak outflow occurs after 8 h and 6 h for 2-, and 100-year storms, respectively. The peak outflow from the RBR is about 0.4 L/s for a 2-year storm, since the soil percolation rate reaches the hydraulic conductivity of loamy sand soil of 30 mm/h. For all storms, it takes almost 72 h for runoff to drain out of the RBR. As shown in Table 4, the peak runoff rate reduction and the runoff volume controlled are about 24% and 36% and 13% and 33% for 2-, and 100-year storms, respectively. Most of the runoff bypasses the catch basin, resulting in relatively low runoff control performance.

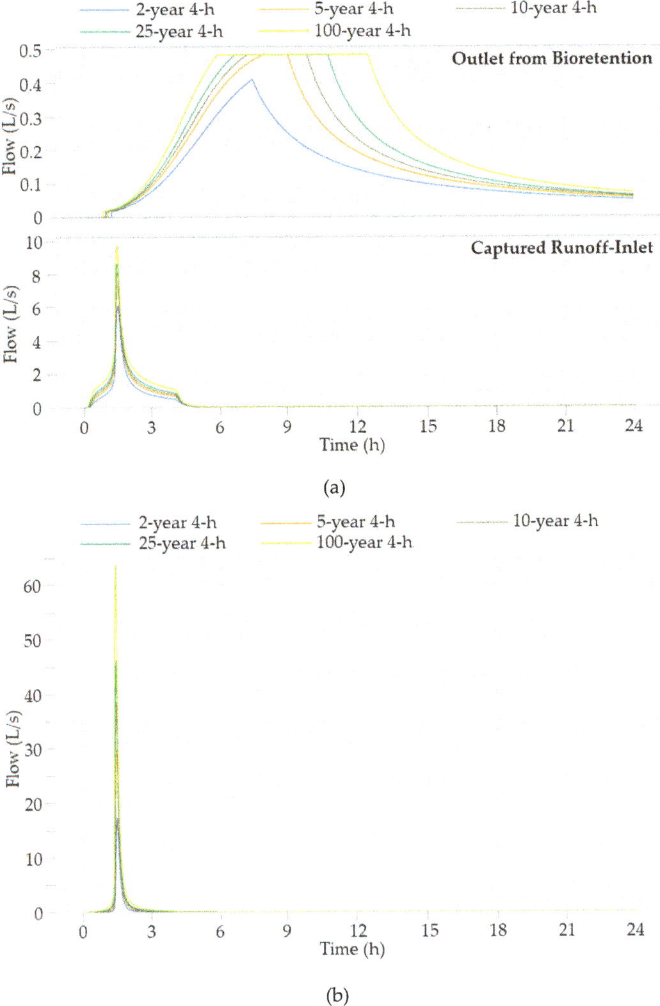

Figure 8. Hydrographs of a single and twin horizontal bar/fishbone catch basin; (**a**) inlet and outlet hydrographs; and (**b**) bypassed hydrographs.

Table 4. Runoff control performances of single & twin horizontal bar/fishbone catch basin.

Single & Twin Horizontal Bar/Fishbone Catch Basins	2-Year	5-Year	10-Year	25-Year	100-Year
Peak Runoff Rate Reduction (%)	24.3	20.2	17.8	16.6	12.6
Runoff Volume Controlled (%)	35.6	32.8	32.2	32.1	32.8

3.1.2. OPSD400.01 Catch Basins

Figure 9 shows the inlet and outlet hydrographs of the RBR and the bypassed hydrograph at the catch basin for 4-h design storms of 2 to 100 year return periods. The bypassed hydrographs (Figure 9b) of 2-, and 100-year storms have peak flow rates 6.5 and 13.6 times higher than those of the inlet hydrographs (Figure 9a). The peak runoff captured by the catch basin occurs after almost 90 min for all storms while the peak outflow occurs after 5.5 h and 8 h for 2-, and 100-year storms, respectively. For this catch basin, the soil percolation rate does not reach the hydraulic conductivity of loamy sand soil. As shown in Table 5, the peak runoff rate reduction and the runoff volume controlled are about 13% and 43% and 7% and 34% for 2-, and 100-year storms, respectively. Most of the runoff bypasses the catch basin, resulting in relatively low runoff control performance.

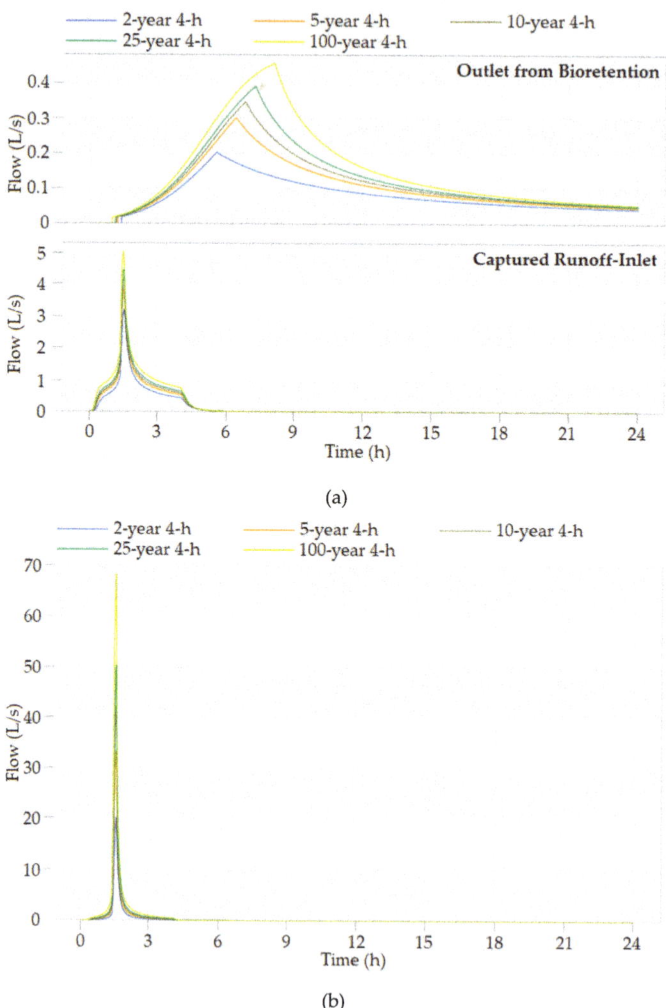

Figure 9. Hydrographs of an OPSD400.01 catch basin (**a**) inlet and outlet hydrographs; and (**b**) bypassed hydrographs.

Table 5. Runoff control performances of OPSD400.01 catch basin.

OPSD400.01 Catch Basin	2-Year	5-Year	10-Year	25-Year	100-Year
Peak Runoff Rate Reduction (%)	13.4	10.1	9.1	8.2	7.1
Runoff Volume Controlled (%)	42.8	38.8	37.2	35.8	33.7

3.1.3. Single and Twin Grid Catch Basins

Figure 10 shows the inlet and outlet hydrographs of the RBR and the bypassed hydrograph at the catch basin for 4-h design storms of 2 to 100 year return periods. The bypassed hydrographs (Figure 10b) of 2-, and 100-year storms have peak flow rates 3.2 and 7.3 times higher than those of the inlet hydrographs (Figure 10a). The results of this type are similar to that of the horizontal bar/fishbone catch basins. The maximum peak outflow is 0.48 L/s when the soil percolation rate reaches the hydraulic conductivity of loamy sand soil. As shown in Table 6, the peak runoff rate reduction and the runoff volume controlled are about 22% and 36% and 13% and 32% for 2-, and 100- year storms, respectively. Most of the runoff bypasses the catch basin, resulting in relatively low runoff control performance.

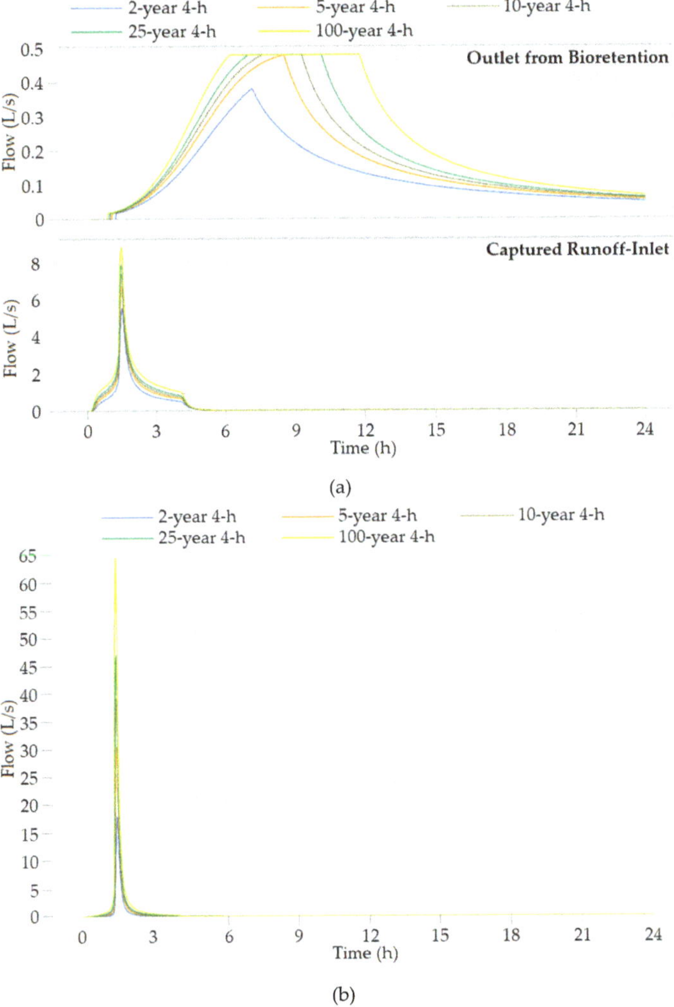

Figure 10. Hydrographs of a single and twin grid catch basin (**a**) inlet and outlet hydrographs; and (**b**) bypassed hydrographs.

Table 6. Runoff control performances of single and twin grid catch basin.

Single & Twin Grid Catch Basins	2-Year	5-Year	10-Year	25-Year	100-Year
Peak Runoff Rate Reduction (%)	22.2	17.8	16.1	14.2	12.5
Runoff Volume Controlled (%)	36.4	33.2	32.5	32	32.3

3.1.4. Single and Twin Honeycomb Catch Basins

Figure 11 shows the inlet and outlet hydrographs of the RBR and the bypassed hydrograph at the catch basin for 4-h design storms of 2- to 100-year return periods. The bypassed hydrographs (Figure 11b) of 2-, and 100-year storms have peak flow rates 6 and 15 times higher than those of the inlet hydrographs (Figure 11a). This honeycomb catch basin has a high capacity for capturing runoff compared to that of horizontal bar and grid catch basins. The peak outflows range from 0.15 to 0.45 L/s. As shown in Table 7, the peak runoff rate reduction and the runoff volume controlled are about 12% and 44% and 6% and 35% for 2-, and 100-year storms, respectively. Most of the runoff bypasses the catch basin, resulting in relatively low runoff control performance.

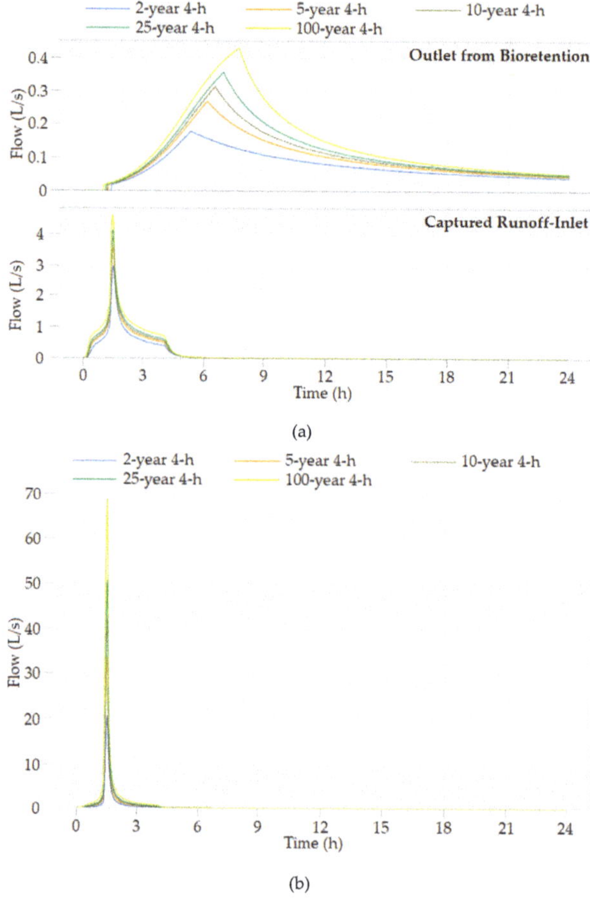

Figure 11. Hydrographs of a single and twin honeycomb catch basin; (**a**) inlet and outlet hydrographs; and (**b**) bypassed hydrographs.

Table 7. Runoff control performances of single and twin honeycomb catch basin.

Single & Twin Honeycomb Catch Basins	2-Year	5-Year	10-Year	25-Year	100-Year
Peak Runoff Rate Reduction (%)	12.3	9.6	7.9	7.6	6.3
Runoff Volume Controlled (%)	43.9	39.9	38.3	36.8	34.5

3.1.5. Performance Comparison of All Catch Basins

Among the four types of catch basins, the single and twin horizontal bar/fishbone catch basin performs the best with the highest peak flow reduction, while the single and twin honeycomb catch basin performs the best with the highest runoff volume controlled. By comparing the inlet and outlet hydrographs of horizontal bar/fishbone catch basin (Figure 8) with those of the honeycomb catch basin (Figure 11), it is clear that the peak inlet flow (6 L/s for a 2-year storm) of the fishbone catch basin is greater than that (3 L/s for a 2-year storm) of the honeycomb catch basin, resulting in higher peak flow reduction. Fishbone catch basin has a higher interception capacity (e.g., 0.014 m^3/s at 0.065 m head) than that (0.005 m^3/s at 0.065 m head) of the honeycomb catch basin [65]. On the other hand, the outlet hydrograph of the fishbone catch basin has larger volume (Figure 8) than that (Figure 11) of the honeycomb catch basin, resulting in lower runoff volume controlled. Nevertheless, none of the catch basins can intercept sufficient runoff to the RBR, resulting in runoff control performance of the RBR less than 45% for a 2-year storm (Tables 4–7).

3.2. *Effect of Inflow Distribution on the Overall Runoff Control Performance of the RBR*

Effective length of a RBR, defined as the extent at which the captured runoff at a catch basin is distributed along a RBR, affect the overall runoff control performance. It depends upon the inflow rate and the perforation arrangement of the inflow distribution pipe. The calibration of the FLOW-3D model was based on (1) mesh sizes of the pipe and the perforations; (2) physics model coefficients such as the turbulence mixing length parameter; (3) boundary conditions such the pipe alignment and perforation locations; and (4) options for volume-of-fluid methods. An accepted goodness of fit criterion of 15% was also assumed to account for the uncertainty of the experimental pipe inner wall boundary and slope, perforation size, orientation, spacing, and level and flow sensors. As indicated in Figure 12, the measured water profiles along the perforated pipe (with 25 mm sized perforations) are non-uniform and the FLOW-3D model simulated water profiles generally match the laboratory measurements (within 10–15%) over the test flow range of 0.63 to 10.28 L/s. For lower inflows (less than 8.08 L/s), the goodness of fit between the measured and the simulated water profiles are consistent. Under the partially surcharged condition (at 10.28 L/s), there is a clear discrepancy between the measured and modelled water profile near the entrance. Given the focus of this study is the free surface water profiles, the calibration is considered acceptable.

The calibrated FLOW-3D model was used to simulate water profiles along the inflow distribution pipe of the Toronto RBR under different inflow rate conditions. The Toronto RBR's inflow distributed pipe has the following characteristics: 150 mm perforated pipe with a total length of 18 m, 19 sets of 10 mm perforations at a distance of 0.9 m apart. Figure 13 shows the simulated water profiles ranging from 2 to 10 L/s. It is noted that the water profiles at 4 L/s or below are gradually varied and do not distribute over the whole length of the pipe. The critical (minimum) inflow, which distributes fully and non-uniformly over the 18 m was found to be about 6 L/s. Two alternative effective lengths, 12 m (67% of 18 m) and 16 m (88% of 18 m), and the full length (i.e., 100%) of the inflow pipe were investigated for the Toronto RBR using the dual drainage PCSWMM model. Figure 14 shows the inflow and outflow hydrographs of 100% (18 m), 88% (16 m), 67% (12 m) of the RBR. For the 2-year design storm (Figure 14a), the inflow hydrograph was increasingly attenuated as the effective length increases (i.e., 67% > 88% > 100%). However, the effect of different effective lengths on the outflow hydrograph is negligible for the 5-year design storm (Figure 14b). Since the difference between the 16 m and 18 m outflow hydrographs is relatively small, the outflow hydrograph of 12 m (67% of the full length) was used to represent the outflow hydrograph for inflow rate less than 6 L/s. As shown in Figure 15, the blue columns are the times that 100% of the RBR would be utilized. The captured inflow only exceeded 6 L/s 2 times for the 2-year storm and 14 times for the 100-year storm. As a result, the outflow hydrographs for 2-, and 100-year storms were adjusted to 67% effective length for those time periods less than 6 L/s while the rest of the outflow hydrographs were assumed to be 100% full length. For the 2-year storm condition, the peak flow reduction and the runoff volume controlled of RBR with outflow adjustment are about 2% less than those without outflow adjustment. There is negligible difference for the 100-year storm condition.

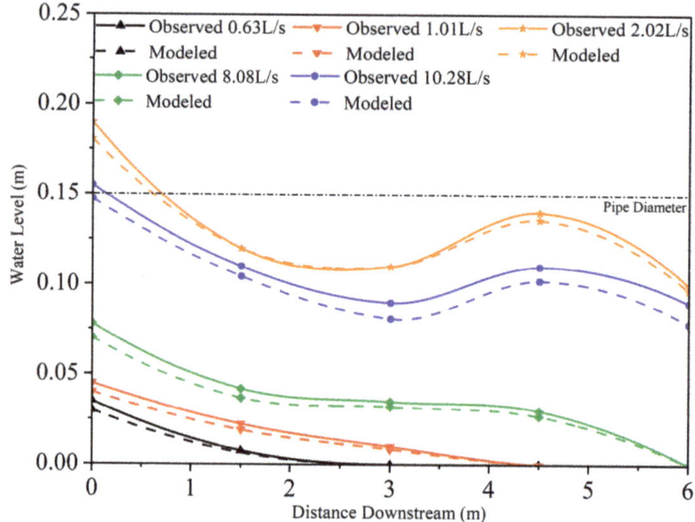

Figure 12. Comparison of the FLOW-3D model simulations with the experimental results.

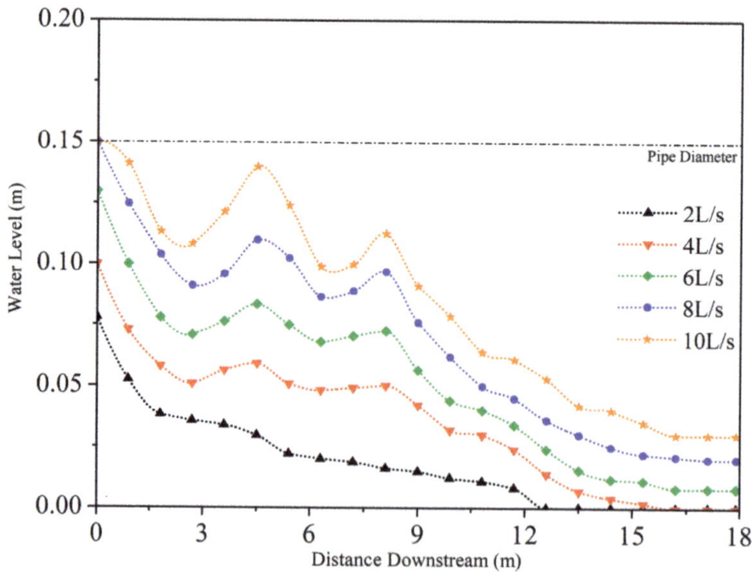

Figure 13. Water profiles for various inflow rates.

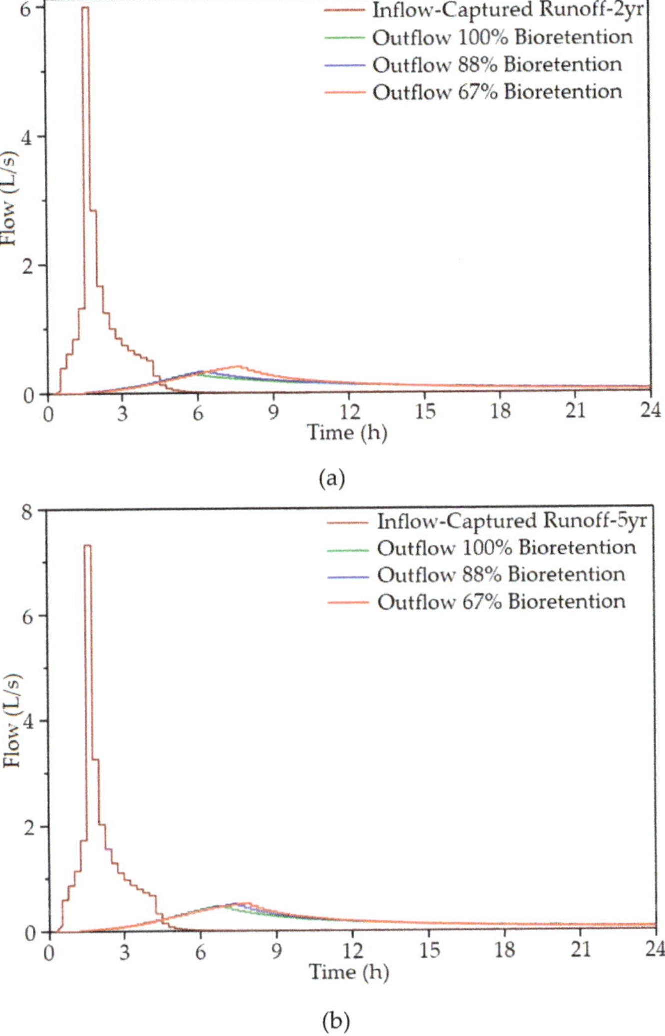

Figure 14. Inflow and outflow hydrographs of RBR for 12, 16, 18 m of length (**a**) 2-year; (**b**) 5-year.

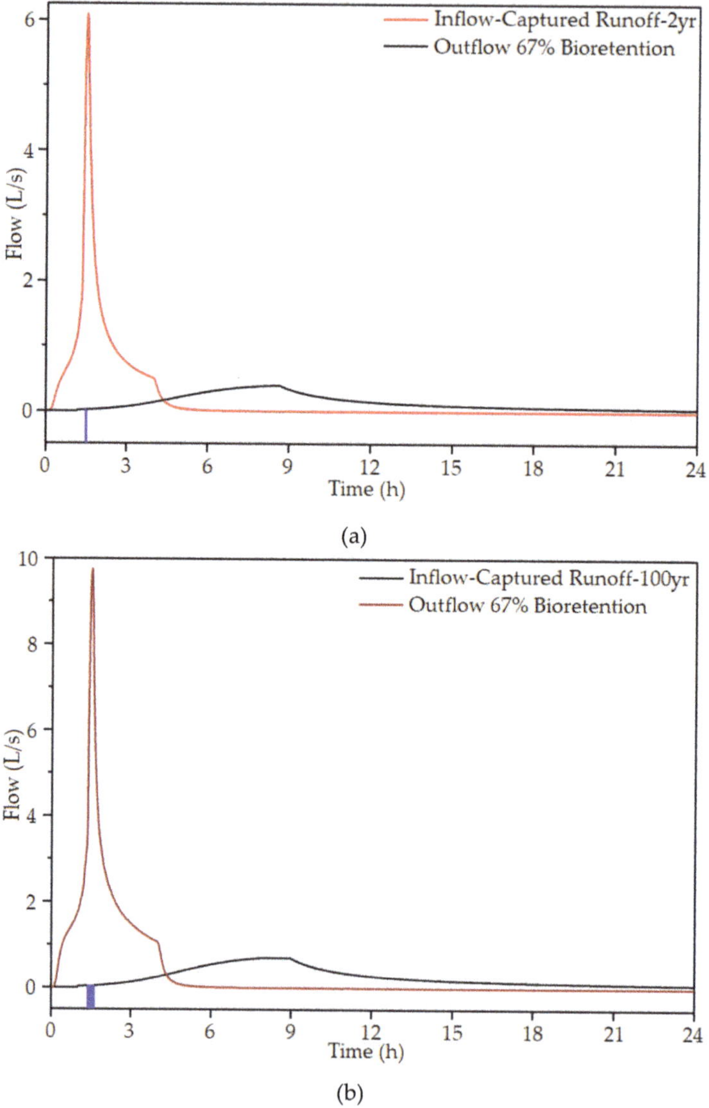

Figure 15. The modified inflow and outflow hydrographs for; (**a**) 2-year storm; (**b**) 100-year storm.

4. Discussion

The first investigation of catch basin effects has demonstated the importance of a catch basin inlet on the overall runoff control performance of a RBR. Although, the number of catch basins along a road are designed to match the drainage capacity of a storm sewer (e.g., runoff of a 2-year storm), an underground RBR is typically aligned with one catch basin. For the four types of catch basins in the Toronto case study, the peak flow rate reduction and the runoff volume controlled for 100-, and 2-year design storms were 12–24%, and 36–44%, respectively because the majority of the runoff bypassed the inlet of the RBR. This case study finding is similar to the intercepting efficiency (35–48%) simulated by a detailed two-dimensional (2D) overland flow simulation model FullSWOF_2D (with grate-inlet and one-dimensional pipe flow) and 20 storm inlet monitored results reported by Li et al. [56].

The second investigation of effective length or effective utilization of a RBR has indicated that water profiles along the inflow distribution pipe can be spatially varied and the hydraulics of inflow distribution pipe affects the runoff control performance of a RBR. However, the inflow pipe characteristics (e.g., number and size of perforations, distance between perforations) of the Toronto RBR do not cause a significant difference of the runoff peak flow reduction and the runoff volume controlled (2% difference for a 2-year storm and negligible difference for a 100-year storm) even the effective length of the RBR was taken into consideration.

5. Conclusions and Recommendations

Based on the aforementioned research, the following conclusions are drawn and recommendations are suggested:

1. No matter how well a RBR is designed (e.g., soil porosity, types of soil, and other outlet features), the inlet dictates its overall runoff control performance. Catch basins are storm inlets and their intercepting capacities vary according to their design (longitudinal and cross slope of road, grate openings and orientation, curb openings). In order to capture more runoff to a RBR, double curb and grate catch basins or multiple inlets should be considered. Recent research studies [49,53,54] also indicate that the traditional hydraulic calculations of catch basin efficiencies may not be adequate. Thus, future investigations of RBR performance should update the traditional inlet efficiency curves of currently available catch basins.

2. The overall runoff control performance of RBR is dependent upon the distribution of inflow over the RBR. In order to fully utilize a RBR, a detailed hydraulic analysis of the inflow distribution pipe, including the perforation size, orientation, spacing, should be performed. Although the case study does not show any significant difference in runoff control performance between the partially and fully utilized RBR, there are situations in which the partially utilized RBR may be important. For instance, if the perforation size is large and their spacing is short, short-curcuiting of flow through the RBR may occur, resulting in large periods of partially utilized RBR and low runoff control performance. Perforated pipes are increasing used as conveyanace and exfiltration components in LIDs, such as grass swale-perforated pipes [66], infiltration trenches [59], exfiltration storage systems [6] and bioretentions [7]. Unfortunately, detailed perforated pipe hydraulic analyses [60,61] are rarely considered in LID modeling. Future research of RBR should not only focus on the associated chemical and biological processes but also the hydraulics of flows at the inlet and perforated distribution pipes.

Author Contributions: Conceptualization, J.L. and D.J.; Methodology, J.L. and D.J.; Software, S.A.; Validation, S.A. and L.C.; Formal Analysis, S.A. and L.C.; Investigation, S.A. and L.C.; Data Curation, S.A. and L.C.; Writing—Original Draft Preparation, J.L.; Writing—Review & Editing, J.L.; Visualization, S.A. and L.C.; Supervision, J.L. and D.J.; Project Administration, J.L.; Funding Acquisition, J.L. All authors have read and agreed to the published version of the manuscript.

Funding: This research was funded by Natural Science and Engineering Research Council, Discovery Grant Number (1-51-50068).

Acknowledgments: The authors would like to acknowledge the support from Natural Science and Engineering Research Council, the City of Toronto, Flow Science Inc., Armtec Inc., Computational Hydraulics International.

Conflicts of Interest: The authors declare no conflict of interest. The founding sponsors had no role in the design of the study; in the collection, analyses, or interpretation of data; in the writing of the manuscript, and in the decision to publish the results.

References

1. Li, J.; Banting, D.; Joksimovic, D.; Fan, C.; Eric, M.; Lawson, S.; Ahmed, N.; Hahn, K.; Mirzajani, M. Evaluation of Low Impact Development Stormwater Technologies for the Uncontrolled Urban Areas in the Lake Simcoe. In *A Report Prepared the Lake Simcoe Region Conservation Authority*; Ryerson University: Toronto, ON, Canada, 2010.
2. Low Impact Development Center, Inc. Introduction to LID: Frequently Asked Questions. In LID Urban Design Tools. 2007. Available online: http://www.lid-stormwater.net/index.html (accessed on 22 February 2019).
3. Transportation Association of Canada. *Geometric Design Guide for Canadian Roads*; Transportation Association of Canada: Ottawa, ON, Canada, 2017.
4. Alam, M.D.; Anwar, A.H.M.; Heitz, A.; Sarker, D.C. Improving Stormwater Quality at Source using Catch Basin Inserts. *J. Environ. Manag.* **2018**, *228*, 393–404. [CrossRef] [PubMed]
5. Alam, M.D.; Anwar, A.H.M.; Sarker, D.C.; Heitz, A.; Rothleitner, C. Characterising Stormwater Gross Pollutants Captured in Catch Basin Inserts. *J. Environ. Manag.* **2017**, *586*, 76–86. [CrossRef] [PubMed]
6. Li, J.; Joksimovic, D.; Tran, J. A Right-of-Way Stormwater Low Impact Development Practice. *J. Water Manag. Model.* **2015**. [CrossRef]
7. Shafique, M. A Review of the Bioretention System for Sustainable Storm Water Management in Urban Areas. *Mater. Geoenvironment* **2016**, *63*, 227–235. [CrossRef]
8. Department of Environmental Resources, The Prince George's County, Maryland. Bioretention Manual. 2007. Available online: https://www.slideshare.net/Sotirakou964/md-prince-georges-county-bioretention-manual (accessed on 2 March 2020).
9. Li, J.; Joksimovic, D.; Chen, L. Stormwater Performance Evaluation of Soil Cells. In *A Report to the City of Toronto*; Ryerson University: Toronto, ON, Canada, 2018.
10. Li, M.-H.; Chan, Y.S.; Kim, M.H.; Chu, K.-H. Assessing Performance of Bioretention Boxes in Hot and Semiarid Regions. *Transp. Res. Rec. J. Transp. Res. Board* **2011**, *2262*, 155–163. [CrossRef]
11. Hart, T.D. Root-enhanced Infiltration in Stormwater Bioretention Facilities in Portland, Oregon. Ph.D. Thesis, Portland State University, Portland, Oregon, OH, USA, 2017.
12. Singh, R.P.; Zhao, F.; Ji, Q.; Saravanna, J.; Fu, D. Design and Performance Characterization of Roadside Bioretention Systems. *Sustainability* **2019**, *11*, 2040. [CrossRef]
13. Soberg, L.C.; Viklander, M.; Blecken, G.-T. Do Salt and Low Temperature Impair Metal Treatment in Stormwater Bioretention Cells with or without a Submerged One. *Sci. Total Environ.* **2017**, *579*, 1588–1599. [CrossRef]
14. Li, Y.; Wen, M.; Li, J.; Chai, B.; Jiang, C. Reduction and Accumulative Characteristics of Dissolved Heavy Metals in Modified Bioretention Media. *Water* **2018**, *10*, 1488. [CrossRef]
15. Mullane, J.M.; Flury, M.; Iqbal, H.; Freeze, P.; Hinman, C.; Cogger, C.G.; Shi, Z. Intermittent Rain Storms Cause Pulses of Nitrogen, Phosphorus, and Copper in Leachate from Compost in Bioretention Systems. *Sci. Total Environ.* **2015**, *537*, 294–303. [CrossRef]
16. Bjorklund, K.; Li, L. Removal of Organic Contaminants in Bioretention Medium Amended with Activated Carbon from Sewage Sludge. *Environ. Sci. Pollut. Res.* **2017**, *24*, 19167–19180. [CrossRef]
17. Blecken, G.-T.; Zinger, Y.; Deletic, A.; Fletcher, T.D.; Hedstrom, A.; Viklander, M. Laboratory Study on Stormwater Biofiltration: Nutrient and Sediment Removal in Cold Temperatures. *J. Hydrol.* **2010**, *394*, 507–514. [CrossRef]
18. Rycewicz-Borecki, M.; McLean, J.E.; Dupont, R.R. Nitrogen and Phosphorus Mass Balance, Retention and Uptake in Six Plant Species Grown in Stormwater Bioretention Mirocosms. *Ecol. Eng.* **2017**, *99*, 409–416. [CrossRef]
19. Endreny, T.; Burke, D.J.; Burchhardt, K.M.; Fabian, M.W.; Kretzer, A.M. Bioretention Column Study of Bacteria Community Response to Salt-Enriched Artificial Stormwater. *J. Environ. Qual* **2012**, *41*, 1951–1959. [CrossRef]
20. Denich, C.; Bradford, A.; Drake, J. Bioretention: Assessing Effects of Winter Salt and Aggregate Application on Plant Health, Media Clogging and Effluent Quality. *Water Qual. Res. J. Can.* **2013**, *48*, 387–399. [CrossRef]
21. Taylor, A.; Wetzel, J.; Mudrock, E.; King, K.; Cameron, J.; Davis, J.; McIntyre, J. Engineering Analysis of Plant and Fungal Contributions to Bioretention Performance. *Water* **2018**, *10*, 1226. [CrossRef]

22. Shetterly, B.J. Soil Phosphorus Characterization and Vulnerability to Release in Urban Stormwater Bioretention Facilities. Master's Thesis, Portland State University, Portland, Oregon, OH, USA, 2018.

23. Luell, S.K.; Hunt, W.F.; Winston, R.J. Evaluation of Undersized Bioretention Stormwater Control Measures for Treatment of Highway Bridge Deck Runoff. *Water Sci. Technol.* **2011**, *64*, 974–979. [CrossRef]

24. Lacy, S. The Fate of Heavy Metals in Highway Stormwater Runoff: The Characterization of a Bioretention Basin in the Midwest. Master's Thesis, University of Kansas, Lawrence, KS, USA, 2007.

25. Muthanna, T.M.; Viklander, M.; Blecken, G.; Thorolfsson, S.T. Snowmelt Pollutant Removal in Bioretention Areas. *Water Res.* **2007**, *41*, 4061–4072. [CrossRef]

26. Kluge, B.; Markert, A.; Facklam, M.; Sommer, H.; Kaiser, M.; Pallasch, M.; Wessolek, G. Metal Accumulation and Hydraulic Performance of Bioretention Systems after Long-term Operation. *J. Soils Sediment* **2018**, *18*, 431–441. [CrossRef]

27. Chen, X.; Peltier, E.; Sturm, B.S.M.; Young, C.B. Nitrogen Removal and Nitrifying and Denitrifying Bacteria Quantification in a Stormwater Bioretention System. *Water Res.* **2013**, *47*, 1691–1700. [CrossRef]

28. Shrestha, P.; Hurley, S.E.; Wemple, B.C. Effects of Different Soil Media, Vegetation, and Hydrologic Treatments on Nutrient and Sediment Removal in Roadside Bioretention Systems. *Ecol. Eng.* **2018**, *112*, 116–131. [CrossRef]

29. Grover, S.P.P.; Cohan, A.; Chan, H.S.; Livesley, S.J.; Beringer, J.; Daly, E. Occasional Large Emissions of Nitrous Oxide and Methane Observed in Stormwater Biofiltration Systems. *Sci. Total Environ.* **2013**, *465*, 64–71. [CrossRef]

30. Shrestha, P.; Hurley, S.E.; Adair, E.C. Soil Media CO_2 and N_2O Fluxes Dynamics from Sand-Based Roadside Bioretention Systems. *Water* **2018**, *10*, 185. [CrossRef]

31. Kansom, T.; Tedoldi, D.; Gromaire, M.C.; Ramier, D.; Saad, M.; Chebbo, G. Horizontal and Vertical Variability of Soil Hydraulic Properties in Roadside Sustainable Drainage Systems (SuDS)-Nature and Implications for Hydrological Performance Evaluation. *Water* **2018**, *10*, 987. [CrossRef]

32. Hunt, W.F.; Lord, B.; Loh, B.; Sia, A. *Plant Selection for Bioretention Systems and Stormwater Treatment Practices*; Springer: Singapore, 2015; ISBN 9789812872449.

33. Hartung, E.W. Aging Bioretention Cells: Do They Still Function to Improve Water Quality. Master's Thesis, Kent State University, Kent, OH, USA, 2017.

34. Palhegyi, G.E. Modeling and Sizing Bioretention Using Flow Duration Control. *J. Hydrol. Eng.* **2010**, *15*, 417–425. [CrossRef]

35. Baek, S.-S.; Ligaray, M.; Park, J.-P.; Shin, H.-S.; Kwon, Y.S.; Brascher, J.T.; Cho, K.H. Developing a Hydrological Simulation Tool to Design Bioretention in a Watershed. *Env. Model. Softw.* **2017**, 1–12. [CrossRef]

36. Zhang, K.; Chui, M.T.F. A Review on Implementing Infiltration-Based Green Infrastructure in Shallow Groundwater Environments: Challenges, Approaches, and Progress. *J. Hydrol.* **2019**, *579*, 1–15. [CrossRef]

37. Roy-Poirier, A.; Champagne, P.; Filion, Y. Review of Bioretention System Research and Design: Past, Present, and Future. *J. Env. Engrg. ASCE* **2010**, *136*, 878–889. [CrossRef]

38. Gulbaz, S.; Kazezyilmaz-Alhan, C.M. An Evaluation of Hydrologic Modeling Performance of EPA SWMM for Bioretention. *Water Sci. Technol.* **2017**, *76*, 3035–3043. [CrossRef]

39. Gao, J.; Pan, J.; Hu, N.; Xie, C. Hydrologic Performance of Bioretention in an Expressway Service Area. *Water Sci. Technol.* **2018**, *77*, 1829–1837. [CrossRef]

40. Brown, R.A. Evaluation of Bioretention Hydrology and Pollutant Removal in the Upper Coastal Plain of North Carolina with Development of a Bioretention Modeling Application in DRAINMOD. Ph.D. Thesis, North Carolina State University, Raleigh, NC, USA, 2011.

41. He, Z.; Davis, A. Process Modeling of Storm-Water Flow in a Bioretention Cell. *J. Irrig. Drain. Eng. ASCE* **2011**, *137*, 121–131. [CrossRef]

42. Meng, Y.; Wang, H.; Chen, J.; Zhang, S. Modeling Hydrology of a Singel Bioretention System with HYDRUS-1D. *Sci. World J.* **2014**, 1–10. [CrossRef]

43. Stewart, R.D.; Lee, J.G.; Shuster, W.D.; Darner, R.A. Modelling Hydrological Response to a Fully-Monitored Urban Bioretention Cell. *Hydrol. Process.* **2017**, *31*, 4626–4638. [CrossRef]

44. Wang, J.; Chua, L.H.C.; Shanahan, P. Hydrological Modeling and Field Validation of a Bioretention Basin. *J. Environ. Manag.* **2019**, *240*, 149–159. [CrossRef] [PubMed]

45. Li, W.S.; Sorteberg, K.K.; Geyer, J.C. Hydraulic Behavior of Storm-Water Inlets: II. Flow into Curb-Opening Inlets. *Sew. Ind. Wastes* **1951**, *23*, 722–7398.

46. Li, W.-H.; Goodell, B.C.; Geyer, J.C. Hydraulic Behavior of Storm-Water Inlets: III. Flow into Deflector Inlets. *Sew. Ind. Wastes* **1954**, *26*, 836–842.

47. Bock, P.; Li, W.-H.; Geyer, J.C. Hydraulic Behavior of Storm-Water Inlets: V. A Simplified Method of Determining Capacities of Single and Multiple Inlets. *Sew. Ind. Wastes* **1956**, *28*, 774–784.

48. Argue, J.R.; Pezzaniti, D. How Reliable are Inlet (Hydraulic) Models at Representing Stormwater Flow. *Sci. Total Environ.* **1996**, *180/190*, 355–359. [CrossRef]

49. Guo, J.C.Y.; MacKenzie, K.A.; Mommandi, A. Design of Street Sump Inlet. *J. Hydraul. Eng. ASCE* **2009**, *135*, 1000–1004. [CrossRef]

50. Gomez, M.; Recasens, J.; Russo, B.; Martinez-Gomariz, E. Assessment of Inlet Efficiency through a 3D Simulation: Numerical and Experimental Comparison. *Water Sci. Technol.* **2016**, *74*, 1926–1935. [CrossRef]

51. Tang, Y.; Zhu, D.; Rajaratnam, N.; Duin, B.V. Experimental Study of Hydraulics and Sediment Capture Efficiency in Catchbasins. *Water Sci. Technolo.* **2016**, *74*, 2717–2726. [CrossRef]

52. Veerappan, R.; Le, J. Hydraulic Efficiency of Road Drainage Inlets for Storm Drainage System under Clogging Effect. In Proceedings of the Proc. 5th International Conference on Flood Risk Management and Response, San Servolo, Italy, 29 June–1 July 2016; pp. 271–281.

53. Schalla, F.E.; Ashraf, M.; Barrett, M.E.; Hodges, B.R. Limitations of Traditional Capacity Equations for Long Curb Inlets. *Transp. Res. Rec. J. Transp. Res. Board* **2017**, *2638*, 97–103. [CrossRef]

54. Rubinato, M.; Lee, S.; Martins, R.; Shucksmith, J.D. Surface to Sewer Flow Exchange through Circular Inlets during Urban Flood Conditions. *J. Hydroinformatics* **2018**, *20*, 564–576. [CrossRef]

55. Virahsawmy, H.K.; Stewardson, M.J.; Vietz, G.; Fletcher, T.D. Factors that Affect the Hydraulic Performance of Raingardens: Implications for Design and Maintenance. *Water Sci. Technolo.* **2014**, *69*, 982–988. [CrossRef]

56. Li, X.; Fang, X.; Gong, Y.; Li, J.; Wang, J.; Chen, G.; Li, M.-H. Evaluating the Road-Bioretention Strip System from a Hydraulic Perspective—Case Studies. *Water* **2018**, *10*, 1788. [CrossRef]

57. SWMM. Available online: https://www.epa.gov/water-research/storm-water-management-model-swmm (accessed on 2 March 2020).

58. PCSWMM. Available online: https://www.pcswmm.com/ (accessed on 10 April 2020).

59. Duchene, M.; McBean, E.A. Discharge Characteristics of Perforated Pipe for Use in Infiltration Trenchces. *Water Res. Bull. AWWA* **1992**, *28*, 517–524. [CrossRef]

60. Afrin, T.; Khan, A.A.; Kaye, N.B.; Testik, F.Y. Numerical Model for the Hydraulic Performance of Perforated Pipe Underdrains Surrounded by Loose Aggregate. *J. Hyraulic. Eng. ASCE* **2016**, *142*, 1–10. [CrossRef]

61. Liu, H.; Zong, Q.; Lv, H.; Jin, J. Analytical Equation for Outflow along the Flow in a Perforated Fluid Distribution Pipe. *PLoS ONE* **2017**, *12*, 1–18. [CrossRef]

62. Qin, Z.; Liu, H.; Wang, Y. Empirical and Quantitative Study of the Velocity Distribution Index of the Perforated Pipe Outflowing Along a Pipeline. *Flow Meas. Instrum.* **2017**, *58*, 46–51. [CrossRef]

63. Paul, U.; Karpf, C.; Schalk, T. Hydraulic Simulation of Perforated Pipe Systems Feeding Vertical Flow Constructed Wetlands. *Water Sci. Technolo.* **2018**, *77*, 1431–1440. [CrossRef]

64. FLOW-3D. Available online: https://www.flow3d.com/ (accessed on 2 March 2020).

65. City of Toronto InfoWorks CS Basement Flooding Model Studies Guideline. Available online: https://wx.toronto.ca/inter/pmmd/callawards.nsf/0/85258049005DEA63852580BB005C66C0/$file/9117-17-7015%20Addendum%203%20Attachment%20-%20Modelling%20Guidelines_October%202014.pdf (accessed on 13 April 2020).

66. Abida, H.; Sabourin, J.F.; Ellouze, M. ANSWAPPS: Model for the Analysis of Grass Swale-Perforated Pipe Systems. *J. Irrig. Drain. Eng. ASCE* **2007**, *133*, 211–221. [CrossRef]

Article

The Effect of Climate Change and Urbanization on the Demand for Low Impact Development for Three Canadian Cities

Sarah Kaykhosravi [1], Usman T. Khan [1,*] and Mojgan A. Jadidi [2]

[1]   Department of Civil Engineering, Lassonde School of Engineering, York University, Toronto, ON M3J 1P3, Canada; saherehk@yorku.ca

[2]   Geomatics Engineering, Department of Earth & Space Science & Engineering, Lassonde School of Engineering, York University, Toronto, ON M3J 1P3, Canada; mjadidi@yorku.ca

*   Correspondence: usman.khan@lassonde.yorku.ca; Tel.: +001-41-6736-2100 (ext. 55890)

Received: 20 March 2020; Accepted: 28 April 2020; Published: 30 April 2020

Abstract: Climate change and urbanization are increasing the intensity and frequency of floods in urban areas. Low Impact Development (LID) is a technique which attenuates runoff and manages urban flooding. However, the impact of climate change and urbanization on the demand or need for LID in cities for both current and future conditions is not known. The primary goal of this research was to evaluate the demand for LID under different climate change and urban growth scenarios based on a physical-based geospatial framework called the hydrological-hydraulic index (HHI). To do this, 12 scenarios considering four climate change and three urbanization conditions were developed. The HHI for three cities in Canada (Toronto, Montreal, and Vancouver) were estimated, evaluated, and compared for these scenarios. The results show that both urbanization and climate change increase the demand for LID. The contribution of climate change and urbanization on LID demand, measured using HHI, varies for each city: in Toronto and Montreal, high rainfall intensity and low permeability mean that climate change is dominant, whereas, in Vancouver, both climate change and urbanization have a similar impact on LID demand. Toronto and Montreal also have a higher overall demand for LID and the rate of increase in demand is higher over the study period. The results of this study provide us with a comprehensive understanding of the effect of climate and urbanization on the demand for LID, which can be used for flood management, urban planning, and sustainable development of cities.

Keywords: climate change; urbanization; low impact development; stormwater; urban runoff; Toronto; Montreal; Vancouver; flooding; geospatial modeling

---

1. Introduction

Floods are a major growing natural hazard that causes the loss of human lives and properties. The proportion of flood occurrence was the highest of all types of natural disasters from 1995 to 2015 [1]. The frequency of flooding has increased during the last two decades (1995 to 2015) [2]. During the same period, floods caused 157,000 fatalities globally (accounting for 26% of all weather-related disasters). In addition, it affected the quality of life of 2.3 billion people, which is the highest share (56%) among weather-related disasters [1]. In terms of economic losses, based on UNISDR (2016), between 2006 and 2016, the average annual costs associated with flooding were about 50 billion USD. This ranks first among all natural disasters. These damages include all types of regions (i.e., land uses) including urban areas [3].

There exist several key drivers of urban flood risk such as urbanization, urban sprawl, increasingly unresponsive engineering, global climate change, and fluvial and pluvial changes [4]. Urbanization causes changes to terrains, slopes, soil type, and vegetative cover, which result in a change of the hydrologic behavior of urbanized areas [5]. Among these changes, the expansion of impervious surfaces dramatically increases the volume of surface runoff [6–9]. The expansion of impervious surfaces can occur in two contexts: (1) within a constant area (by an increase of the imperviousness densification within a fixed area); and (2) by increasing the area of the city (expansion of impervious area). However, in our research, we only considered the first context. On the other hand, due to a higher population density and valuable properties in urban areas, the influence of floods on these environments is intensive and excessively costly. Thus, urban areas are more vulnerable to the effect of floods, and, by the growth of the urban population, therefore, vulnerability and exposure are also significantly higher [2]. Climate change and the increase of rainfall intensity have also increased flood risk associated with the frequency of the occurrence of the flood [2], and this trend is expected to continue in many regions.

To cope with this, the traditional and most common approach is using a stormwater collection network (SCN). The objective of SCN is to collect and rapidly convey surface runoff to a predetermined outfall (e.g., receiving water bodies such as rivers, lakes, or ponds) [10]. However, several issues are associated with SCN. SCN is designed for a specific rainfall event; thus, there is an evident limitation of the capacity to manage extreme flood events which are likely due to climate change. This issue is exacerbated by the fact that existing SCNs are aging and performing at a lower capacity. Besides, SCN collects polluted runoff from urban surfaces, which causes environmental damages and financial losses. These losses are either due to contamination of receiving waters or treating the collected runoff at stormwater treatment plants. Even though using SCN is a well-known and well-documented method [11,12], it has several issues, which necessitate a different approach to managing runoff in urban areas.

The smart and sustainable urban growth approach commonly known as Low Impact Development (LID) is an innovative solution to complement SCN, which is increasingly being adopted in cities around the world [13–17]. LID can be incorporated with existing SCN to increase the capacity of the SCN and, thus, improve their performance. LID allows for reducing the risk of floods by reducing urban runoff [14,17–22]. This is achieved by reducing imperviousness, conserving natural resources and ecosystems, maintaining natural drainage courses, reducing the use of pipes, and minimizing clearing and grading during the development process. Structural techniques such as bioretention cells, rain gardens, green roofs, and permeable pavements are also used to increase infiltration, thereby decreasing the amount of runoff and delaying the time to peak [15,23,24].

Demand for LID in different locations varies based on different factors such as land cover type, topography, etc. [25–27]. From a hydrological-hydraulic point of view, this demand is associated with the extent of the runoff generated in each site. That is, the higher is the generated runoff, the higher is the demand for capturing the runoff at the source (the site) [18,25]. The change in rainfall intensity and depth (caused by climate change) and land cover type (caused by urbanization) are the two main factors that affect the extent of this demand. In a specific site, if the rainfall or the land cover changes, the demand for LID will change accordingly [25]. This is due to the change in the volume of runoff generated at that site, which is a function of both climate and land cover.

At the city-scale additional complexity is introduced due to numerous LID sites, constant urban change (and growth), global climate change, and the non-uniform spatial distribution of land cover and rainfall [2,28–32]. Thus, it is extremely important to be able to quantify this complexity to attain insight into how changes in climate and urban land cover influence the demand for LID. Such information is useful for urban planning, designing multifunctional urban infrastructure, policymaking regarding urban growth, and the optimal assignment of financial resources for runoff and flood risk management. Therefore, the objective of our study was to investigate the effect of climate change and urbanization on the demand for LID.

The results of such an investigation will provide us with an insight into the effect of climate and urbanization on the demand for LID. This article holistically covers the scenarios under which either only one change takes place (i.e., climate or urbanization) or both (i.e., climate and urbanization), and discusses the causes and effects of the future demand for LID for three selected Canadian cities. It also discusses the contribution of climate change and urbanization factors on the change of hydrological and hydraulic processes. The manuscript is organized as follows. Section 2 presents the methods including the study area, the geospatial data, and the development of climate change and urban growth scenarios. Sections 3 and 4 include the results and discussion, respectively, followed with the conclusion in Section 5.

2. Methods

In our study, we used a physical-based geospatial index called the hydrological-hydraulic index (HHI) [25]. We estimated HHI for three Canadian cities (Toronto, Montreal, and Vancouver) using the approach proposed Kaykhosravi et al. [25]. The HHI was generated for these cities under various climate and urbanization scenarios. For urbanization, we developed different land cover scenarios to represent different degrees of future urbanization, by modifying existing pervious areas to impervious areas. For the climate change scenarios, we created spatial rainfall distribution maps based on the projected Intensity–Duration–Frequency (IDF) curves for each city. These maps included historical data as well as three future climate change scenarios. The urbanization and climate change scenarios were integrated and used to estimate, evaluate, and compare the HHI (a measure of LID demand [25]) for each city.

2.1. Description of the Three Canadian Cities

The study area in this research includes three metropolitan Canadian cities: Toronto, Montreal, and Vancouver. The location of all cities and their administrative border are presented in Figure 1. The figure illustrates the current status of land cover in all cities, categorized into three classes: impervious lands, pervious lands, and water bodies.

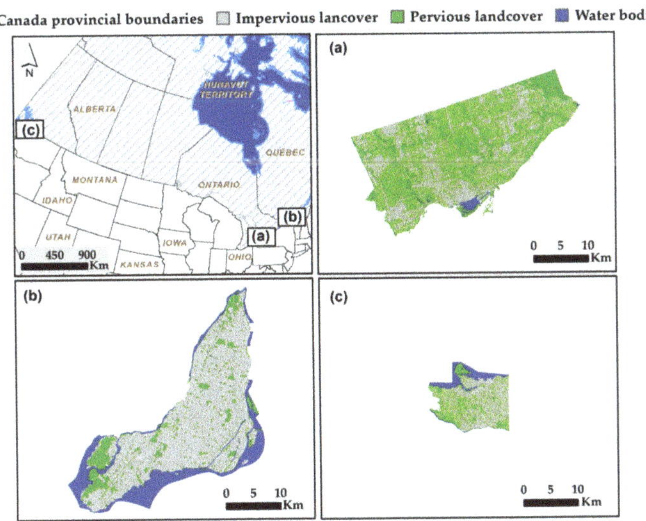

Figure 1. The location of the three case-study cities within Canada showing three land cover categories: (a) Toronto; (b) Montreal; and (c) Vancouver.

The area of Toronto is 633.5 km^2 and the population of the city in 2019 was 2.96 million [33]. The average annual precipitation of Toronto for the 1951–1980 period is 814 mm [34] which is the lowest precipitation among the three cities. The pervious to impervious ratio (PIR) (ratio of areas excluding water bodies) is about 1.17 based on the available land cover data (from 2007) meaning that land cover in Toronto is dominated by pervious land cover.

The area of Montreal is 624.4 km^2, slightly lower than Toronto, and the population of the city in 2016 (based on the most recent census data) is 1.96 million [35]. The average annual precipitation for the 1951–1980 period is 1028 mm [34], which is slightly higher than in Toronto. The PIR for Montreal is 0.21 which is the lowest among all three cities, meaning that the city is highly urbanized or, in other words, dominated by impervious land cover.

Vancouver area is 136.5 km^2 and the population of the city in 2016 (based on the most recent census year) is 2.55 million [36]. The average annual precipitation for the 1951–1980 period is 1569 mm [34], which shows the city has the highest precipitation depth among the three cities. The PIR for Vancouver according to the available land cover (from 2014) is 0.64, which is higher than Montreal but lower than Toronto. In addition, this indicates that the land cover of Vancouver is largely dominated by impervious land cover.

2.2. Geospatial Data Used for Generating the HHI

Our analysis in this study was conducted based on geospatial analysis. The HHI model was developed using Python and the ArcPy package was used to perform the analysis. The data used to estimate HHI included:

- Meteorological station data: To estimate rainfall intensity
- Surficial geology: To estimate the depth to restrictive layer and the hydraulic conductivity
- Groundwater tables: To estimate the depth to groundwater
- Digital Elevation Model (DEM) data: To estimate the slope

All geospatial data used in this study are publicly available and were downloaded from various open data sources [37–43]. All datasets were converted into raster format with a ground resolution of 5 m × 5 m (to align with the digital elevation model of Toronto). This resolution was used as the reference resolution for this research.

The raster maps that were derived from these data were overlaid using Equation (1) [25] to generate the HHI. HHI ranks the cells (i.e., the pixels of the raster data) by their runoff generation potential. Thus, cells with higher HHI generate higher volume and peak runoff flow rate. Ranking the cells based on HHI values highlights the cells that are the main source of runoff generation. On this basis, we can target the high HHI cells for LID implementation, and thereby effectively maximize runoff reduction [25]. Since the demand for implementation of LID corresponds to the HHI values, site with higher HHI values are considered as sites with higher demand for LID [25]. The HHI index accounts for seven variables, as presented in Equation (1).

$$\text{HHI}_j = \begin{cases} \left[\left[R_j - \left((K_s)_j \cap (K_i)_j \right) \right] \cap \left(n_j \times \left((D_g)_j \cap (D_r)_j \right) \right) \right] \times \left[1 + (\tan S)^{0.385} \right] & \text{for } R_j > Ks_j \\ 0 & \text{for } R_j \leq Ks_j \end{cases} \tag{1}$$

In Equation (1), j is the cell number; HHI_j is HHI at cell j; R_j is rainfall intensity (mm/h) at cell j; $(K_s)_j$ is saturated hydraulic conductivity of the surficial soil (mm/h) at cell j; $(K_i)_j$ is the hydraulic conductivity of impervious areas (e.g., roads, parking lots, and building rooftops) at cell j, which is set equal to zero for impervious cells; n_j is the soil porosity at cell j; $(D_g)_j$ and $(D_r)_j$ are, respectively, depth to groundwater (mm) and depth to the first restrictive soil layer (mm) at cell j; and S is the terrain slope (degrees) at cell j [25].

2.3. Developing Scenarios for Investigating the Integrated Effect of Climate Change and Urbanization

2.3.1. Climate Change Scenarios

The pattern of the effective rainfall (rainfall with losses deducted) can have a significant influence on the runoff hydrograph [44]. The pattern is typically determined by selecting the position of the maximum rainfall block (in the hyetograph) and rainfall duration [45]. For hydrological modeling of a catchment, it is important to determine the duration of the rainfall to which flood generation is assumed to occur [46]. Generally, for catchments above 40 km^2, the effective daily rainfall duration is in the order of 12 h (with the time of concentration between 4 and 12 h); for catchments larger than 500 km^2, or slow-reacting catchments, a sub-daily distribution is not needed. Smaller catchments are reactive to short but intensive rainfall events in producing their highest peaks [46]. For example, the recommended duration is 10 min inlet time for small (micro-scale) catchments, whereas the actual time of concentration should be used for larger-scale catchments, according to the Wet Weather Flow Management Guidelines for Toronto [47]. In this study, the catchments were generally small and, therefore, shorter durations (less than the time of concentration) were used for the analysis. However, due to the lack of such fine resolution data for all three cities, a duration of 60 min was used for effective rainfall for each city for the HHI analysis. In addition, a 10-year return period (RP) was selected because this is the most extreme scenario suggested by Canadian guidelines for the design of small wet weather systems including LID. The RPs suggested by these guidelines include 2 [47–49], 5 [48], or 10-year [47,50].

For the projected IDF data, we used the IDF_CC Tool 4.0 (Computerized Tool for the Development of Intensity-Duration-Frequency Curves under Climate Change—Version 4.0) [51]. The available range of IDF data available through this tool are the years 2006–2100. Generating IDF curves based on stationary historical data cannot capture the changing climate conditions; thus, this tool uses two other models: Global Climate Models (GCMs) and Regional Climate Models (RCMs) [52]. Both GCMs and RCMs need spatial and temporal downscaling, which relies on the historically observed data for the time-period of observations [52]. Downscaling links GCM/RCM grid scales and local study areas for the generation of IDF curves under changing climate conditions. For temporal downscaling of precipitation data, a modified version of the equidistant quantile-matching (EQM) method is used in this tool. For spatial downscaling, the tool utilizes a statistical downscaling method. The first future climate that this tool provides is for the period 2006–2036 (referred to as Scenario 2036 in this study), which was selected as our first future scenario [53]. In addition, projections for the period 2036–2066 (referred to as Scenario 2066 in this study) as the mid-term scenario and the period 2066–2096 (referred to as Scenario 2096 in this study) for the long-term period were also selected. Data from the Canadian Climate Model (CanESM 2) were selected from among several available climate models within the tool. The tool generates the IDF for three different values of the Representative Concentration Pathways (RCP) [54] including RCP2.6, RCP4.5, and RCP8.5. The comparison of IDF for these three RCPs for various RPs (2, 5, 10, 25, 50, and 100) provided by the tool shows that the lower the RP the smaller is the difference between IDFs with different RCPs [51]. Thus, for the RPs of 2–10 years, the difference between IDF for different RCPs is almost negligible. Thus, we selected the 10-year RP and the highest RCP (8.5 W/m^2) to derive the projected IDF values for the selected scenarios (2036, 2066, and 2096).

Following the derivation of rainfall intensity using the IDF_CC Tool 4.0 at each meteorological station within or near the cities, the geospatial distribution of rainfall across each city was estimated using the stations listed in Tables 1–3 for each city. This process was as follows: First, using the IDF tables, the rainfall intensities for the 10-year RP and 60 min duration were derived. Second, a geospatial analysis (using Inverse Distance Weighted method, IDW) was used to develop the distribution of the rainfall intensity across the study areas based on the location of the stations. This resulted in 12 different rainfall distribution maps (four for each city for the selected years including historical IDF and projected IDF for Scenarios 2036, 2066, and 2096).

Table 1. Projected rainfall intensity for Toronto for Scenarios 2036, 2066, and 2096 [51,55], 60 min duration, and 10-year RP. The coordinates are in NAD_1983_UTM_Zone_17N.

ID	Station Name	Historical Data Range	Coordinates		Rainfall Intensity (mm/h)			
			X (m)	Y (m)	Historical	2036	2066	2096
1	Oshawa WPCP	1970–2006	674,358.9	4,859,722	32.17	35.00	35.45	42.12
2	Toronto City	1940–2007	628,988.2	4,836,464	39.12	43.54	43.59	49.43
3	Toronto-INTL A	1950–2017	610,427.6	4,837,243	38.12	42.06	42.99	47.39
4	Toronto Buttonville A	1986–2007	630,991.4	4,857,614	37.50	39.56	39.94	45.49
5	Oakville Southeast	1965–1976	610,793.8	4,815,031	32.33	35.70	36.48	40.10
6	Etobicoke	1964–1980	618,586.5	4,831,829	36.05	39.77	40.09	45.12
7	Toronto Met Res Stn	1966–1987	616,643.0	4,850,681	36.91	40.62	41.62	45.96

Table 2. Projected rainfall intensity for Montreal for Scenarios 2036, 2066, and 2096, 60 min duration [51], and 10-year RP. The coordinates are in NAD_1983_UTM_Zone_18N.

ID	Station Name	Historical Data Range	Coordinates		Rainfall Intensity (mm/h)			
			X (m)	Y (m)	Historical	2036	2066	2096
1	McGill	1906–1992	609,377.3	5,039,448	42.20	45.70	55.45	55.70
2	Mirabel Int. A.	1976–2008	575,554.5	5,057,840	32.41	33.70	40.93	41.58
3	P. E. T. Int. A.	1943–2014	598,491.9	5,035,934	34.90	36.66	45.37	45.66
4	St Hubert A.	1956–1995	623,396.3	5,041,931	34.60	36.42	44.99	44.81
5	Jean Brebeuf	1969–1984	609,377.3	5,039,448	32.50	33.91	41.70	42.14
6	Lafontaine	1973–1991	609,338.6	5,041,670	32.35	33.46	41.63	41.99
7	Jar Bot	1977–1989	613,143.1	5,047,295	39.35	39.54	50.80	49.64
8	Valleyfield	1986–1998	570,587.7	5,014,450	32.61	34.99	41.20	41.62
9	L'Assomption	1963–2017	621,983.8	5,074,136	36.28	38.29	47.44	47.43

Table 3. Projected rainfall intensity for Vancouver for Scenarios 2036, 2066, and 2096 [51,56], 60 min duration, and 10-year RP. The coordinates are in NAD_1983_UTM_Zone_10N.

ID	Station Name	Historical Data Range	Coordinates		Rainfall Intensity (mm/h)			
			X (m)	Y (m)	Historical	2036	2066	2096
1	UBC	1958–1990	481,806.10	5,455,276	14.35	14.98	15.60	16.84
2	Lynn Creek	1964–1983	497,822.00	5,468,587	19.89	21.05	21.94	23.65
3	Ladner Bchpa	1963–1978	503,651.30	5,436,349	11.02	11.68	11.99	13.04
4	Surrey Kwantlen Park	1962–1999	510,200.90	5,448,586	15.79	16.59	17.27	18.53
5	Buntzen Lake	1969–1983	509,436.00	5,469,707	21.30	22.35	23.43	25.03

2.3.2. Urban Growth Scenarios

The projected land cover of each city was required to investigate the influence of future urbanization on runoff generation potential. To develop these projected scenarios, there were several limitations due to inconsistent and/or complete lack of data for the study periods. To account for the inconsistencies in the land cover categories, we generalized the land covers into three categories: impervious (buildings, roads, parking lots, and other paved surfaces), pervious (tree canopies, parks, golfs, grass shrubs, bare earth, etc.), and water body (lakes, ponds, streams, etc.). For example, in Toronto, all trees (street, park, and forest trees) are considered as one category (tree canopy), whereas, in other two cities, trees are not distinguished from other green spaces, and are part of open spaces or park areas. Thus, due to this inconsistency, comparing the land covers directly for each city is not feasible. Tables 4–6 illustrate these differences for each city. These tables show 8 categories of land cover for Toronto and Montreal and 13 for Vancouver, each with a different taxonomy.

Table 4. Land cover data of Toronto at the year 2007 and its projected changes due to the urban growth for two scenarios.

ID	Land Cover	Area (km^2)	Base	Scenario 1	Scenario 2
1	Buildings	103	impervious	impervious	impervious
2	Roads	79.8	impervious	impervious	impervious
3	Paved surfaces	108	impervious	impervious	impervious
4	Tree canopy	179.2	pervious	pervious	impervious
5	Grass shrub	149.4	pervious	impervious	impervious
6	Agriculture	6.2	pervious	impervious	impervious
7	Bare earth	4.1	pervious	impervious	impervious
8	Water bodies	3.8	-	-	-
	Total	633.5			
	Added impervious		0	82.4 [1]	139 [1]
	Total impervious		290.7	373.1	429.7
	Total pervious		338.9	256.6	200.4
	PIR		1.17	0.69	0.47

[1] Pervious lands within a buffer of 500 m from the streams were not converted to impervious.

Table 5. Land cover data of Montreal at the year 2016 and its projected changes due to the urban growth for two scenarios.

ID	Land Cover	Area (ha)	Base	Scenario 1	Scenario 2
1	Buildings	264.6	impervious	impervious	impervious
2	Roads	103.5	impervious	impervious	impervious
3	Paved surfaces	47	impervious	impervious	impervious
4	Parks and open spaces	41.1	pervious	pervious	impervious
5	Golf	12.5	pervious	impervious	impervious
6	Agriculture	6.1	pervious	impervious	impervious
7	Bare earth	25.3	pervious	impervious	impervious
8	Water bodies	124.5	-	-	-
	Total	624.6			
	Added impervious		0	19.5 [1]	41.1 [1]
	Total impervious		415.1	434.6	456.2
	Total pervious		85.1	65.3	43.8
	PIR		0.21	0.15	0.10

[1] Pervious lands within a buffer of 500 m from the streams were not converted to impervious.

Another limitation was that the future urban growth data (which were used to estimate future urbanization) for the selected years were not available. To address this, we used a systematic urban growth model to generate this data. In the proposed model, we estimated the future state of cells (i.e., locations within each city). Our model is a grid-based discretization method [57] which is different from conventional urban growth models such as cellular automata (CA), land use/transportation (LUT), or agent-based (ABM) models [58]. The selection and priority of the "converted lands" were made based on a combination of three parameters including the hydrological state of the lands (pervious or impervious) considering the land cover (i.e., pervious lands such as bare earth were selected for conversion); the functional importance of the land cover (e.g., tree canopy had priority to bare earth for conversion to impervious cover); and finally the total geometrical area of the selected land cover (e.g., tree canopy in Toronto has a significant share in the area of the city which is greater than the area of all other pervious lands such as grass shrubs, bare earth, etc.). We divided the conversion process into two steps. In Step 1 (Scenario 1, representing medium urbanization), a few land covers are converted to impervious surfaces, whereas in Step 2 (Scenario 2, representing high urbanization) more land covers follow (shown red colored text in Tables 4–6). Subsequently, the corresponding area of conversions is calculated, leading to different values for different cities. The PIR values for all three cities and scenarios are presented in Tables 4–6. Overall, as expected, PIR decreases with urbanization in all cities.

Table 6. Land cover data of Vancouver at the year 2014 [59] and its projected changes due to the urban growth for Scenario 1.

ID	Land Cover	Area (ha)	Base	Scenario 1	Scenario 2
1	Buildings	37.5	impervious	impervious	impervious
2	Other Built	0.1	impervious	impervious	impervious
3	Paved	33.3	impervious	impervious	impervious
4	Coniferous	6	pervious	pervious	impervious
5	Deciduous	21.8	pervious	pervious	impervious
6	Barren	1.9	pervious	pervious	impervious
7	Shrub	0.7	pervious	impervious	impervious
8	Shadow (considered as shrub)	1.5	pervious	pervious	impervious
9	Modified Grass-herb	13	pervious	impervious	impervious
10	Natural Grass-herb	0.2	pervious	impervious	impervious
11	Non-photosynthetic vegetation	0.1	pervious	pervious	impervious
12	Soil	0.3	pervious	pervious	impervious
13	Water	20.1	-	-	-
	Total			136.5	
	Added impervious		0	13.8	34.3 [1]
	Total impervious		71	84.8	105.3
	Total pervious		45.4	31.6	11.1
	PIR		0.64	0.37	0.11

[1] Pervious lands within a buffer of 500 m from the streams were not converted to impervious.

For Toronto, the impervious area increased by 28% and 48% for Scenarios 1 and 2, respectively. For Montreal, the impervious area increased by 5% and 10% for Scenarios 1 and 2, respectively. Lastly, for Vancouver, the impervious area increased by 19% and 48% for Scenarios 1 and 2, respectively. In each of these cases, the increase in impervious area is measured by comparison with the base scenario for each city. The difference among the "added impervious areas" of the cities for the same scenario is due to two reasons: (1) the limited capacity of the cities in terms of available pervious land cover; and (2) the inconsistency in the base maps in terms of the types of land cover (as described above). This limited the consistent conversion for the same scenario for all cities. For example, "grass shrub" in Toronto does not have a similar land cover type in Montreal ("Golf" can be considered a similar class), whereas "shrub", "modified grass herb", or "natural grass herb" can be considered as similar cover types in Vancouver.

In the highly urbanized state (Scenario 2), the PIR is 0.47, 0.10, and 0.11 for Toronto, Montreal, and Vancouver, respectively. These ratios reveal how far the cities are from the highly urbanized state. Both Montreal and Vancouver have similar PIR values for the most critical condition (i.e., Scenario 2) with the lowest pervious land cover, whereas Toronto has a significantly high proportion of pervious land cover. Overall, in Scenario 2, in all three cities, the area of impervious lands dominates the pervious area.

To verify that the urban growth scenarios do not overestimate future urban areas, the computed scenarios described above were compared to a population-based metric. This metric denoted as the Additional Required Impervious Lands (ARIL) is presented in Equations (2) and (3). ARIL estimates how much impervious area is needed on top of the base land cover (i.e., current conditions) to accommodate the projected population in the future. To calculate ARIL, a linear relationship between population growth and impervious area (km^2) was assumed using similar methods proposed in previous studies (e.g., [60,61]). In addition, we assumed that the population distribution over impervious lands is homogenous, the population density in the future remains the same, and the future area of impervious lands is a linear function of the current area and population density. In other words, future urbanization and urban growth will continue in the future as they are currently.

$$\text{PPI (capita/ha)} = P_{base} \text{ (capita)}/Ai_{base} \text{ (km}^2) \tag{2}$$

$$\text{ARIL} = (P_x - P_{base})/\text{PPI} \tag{3}$$

where Ai_{base} is the area of impervious lands at the base year, P_x is the projected population at year x, P_{base} is the population at year b, and PPI is the population per impervious land (Equation (2)).

The projected populations for Toronto, Montreal, and Vancouver are needed to estimate the ARIL (Table 7). The available data for these projections for each city had several inconsistencies creating a challenge in data integration, particularly due to missing data for the three future periods. To overcome this, the missing data were estimated using a linear function to the available data (which included historical population data for each city as well as projected data for certain future years), for each city. The project data are specified in Table 7, which also identifies which data were estimated using the linear function.

Table 7. The projected population for Toronto, Montreal, and Vancouver scenarios 2036, 2066, and 2096 [35,36,62].

		Scenario			
		Base	**2036**	**2066**	**2096**
Population (million)	Toronto	2.40 [2] (year 2007)	3.81 [1]	5.10 [2]	6.47 [2]
	Montreal	1.96 [3] (year 2016)	2.27 [1]	2.70 [2]	3.14 [2]
	Vancouver	2.42 [3] (year 2014)	3.40 [1]	4.35 [2]	5.47 [2]

[1] Based on projected data. [2] Estimated using a linear function. [3] Based on available historical data.

Based on Equations (2) and (3), PPI in 2007 (the base year) for Toronto is estimated as 8255.9 capita/km^2. For Montreal, the PPI calculated in the year 2016, for which both population and land cover data are available, was estimated as 5159.8 capita/km^2. Vancouver had a PPI of 34,084.5 capita/km^2 in year 2016, for which both population and land cover data are available. Using PPI values, the ARIL of each city for all projected years were estimated and are presented in Table 8. We compared the ARIL against the developed scenarios. This revealed that the added impervious lands (by converting pervious lands) in two scenarios are not adequate to accommodate the projected population with an assumption of constant PPI. In all cases, the ARILs in Scenarios 2066 and 2096 are larger than the added impervious area presented in Tables 4–6. From this, we learned that for the future land cover scenarios, only converting the pervious to impervious lands is not sufficient to accommodate such projected populations and the PPI needs to increase. In other words, the current lifestyles will need to be modified in each city to account for the increase in population, e.g., through the vertical development of buildings (i.e., densification) or an increase in population density.

Table 8. Land cover data of Toronto, Montreal, and Vancouver on the base year (2007, 2016, and 2014, respectively for each city) and its projected changes due to the urban growth in years 2030, 2066, and 2096.

		Base	2036	2066	2096
Toronto	ARIL (km^2)	0.0	170.8	327.2	493.1
	Total impervious (km^2)	290.7	461.5	788.6	1281.7
	Total pervious area (km^2)	338.9	168.1	−159.0	−652.1
Montreal	ARIL (km^2)	0	60.08	143.61	228.89
	Total impervious (km^2)	379.86	439.94	583.55	812.44
	Total pervious area (km^2)	120.13	60	−84	−312
Vancouver	ARIL (km^2)	0	28.75	56.62	89.51
	Total impervious (km^2)	71	99.75	156.38	245.89
	Total pervious area (km^2)	45.4	17	−40	−129

2.3.3. Integrated Urban Growth and Climate Change Scenarios

We investigated all possible combinations of our selected urbanization and climate change scenarios. Using this, the influence of the sole change of any scenario as well as the combined change of any land cover and climate scenarios can be analyzed. The developed scenarios are labeled and presented in Table 9.

Table 9. The 12 combinations of climate change and urbanization scenarios.

		Climate Change			
		Base Year	**Sc 2036**	**Sc 2066**	**Sc 2096**
Urbanization Scenarios	Sc00 (base land cover)	C-base-00	C-2036-00	C-2066-00	C-2096-00
	Sc01 (medium urbanized)	C-base-01	C-2036-01	C-2066-01	C-2096-01
	Sc02 (highly urbanized)	C-base-02	C-2036-02	C-2066-02	C-2096-02

3. Results

A brief comparison of climate condition and urbanization under different scenarios is presented in Figures 2 and 3. The mean rainfall intensity of the cities shows that the rainfall intensity of Toronto is slightly higher than that of Montreal and more than double the rainfall intensity of Vancouver. However, Vancouver has a maximum average annual rainfall of 1569 mm, about 90% higher than Toronto and 50% higher than Montreal. Overall, the trend of projected mean rainfall intensity presents an increase in the intensity of rainfall in the future for each city for each future period. However, there is an exception for Toronto in Scenario 2066, where the intensity is shown to remain almost constant compared to the previous scenario (2036). This is similar to Montreal for Scenario 2096 compared to the previous scenario (2066). Note that the variability of rainfall intensity is higher in Montreal compared to the other two cities.

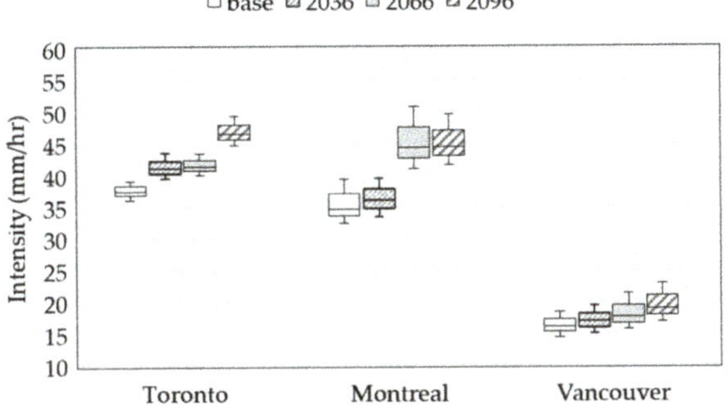

Figure 2. Comparison of rainfall intensity (for the 60 min duration and 10-year RP event) for the three cities in selected years.

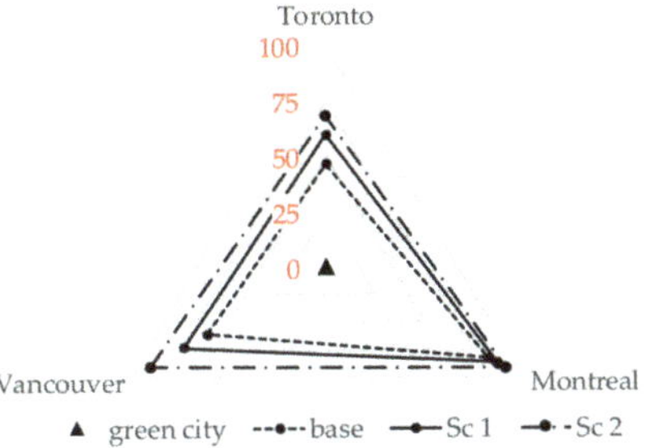

Figure 3. Normalized ratio of impervious area to the city area (water bodies excluded) for Toronto, Montreal, and Vancouver.

Figure 3 presents an overall comparison of all urbanization scenarios for the three cities. If zero impervious surface is defined as the "green city", Figure 3 shows that Toronto (for the base scenario) is the closest to the "green city" and Montreal is the most urbanized city (closest to 100% impervious cover). As expected, the two projected scenarios (Sc1 and Sc2 representing medium and high urbanization) move each city further away from the "green city".

For the HHI, first, a general comparison of the three cities is presented in Figure 4 which shows the current state (C-base-00, i.e., the base climate and the base urbanization scenarios) as well as the worst-case scenario (C-2096-02, i.e., Scenario 2096 climate and high urbanization, Scenario 2). Note that, as identified above, the C-base-00 comparison has some limitations: the base land cover for these cities is not for the same year, while the rainfall intensity is based on available historical data (which differs for each city).

This comparison was made based on the mean and frequency of HHI in each city. Figure 4 shows the normalized histogram of HHI for each city for the two scenarios listed above; a low HHI value represents a lower demand for LID, and a higher value represents a higher demand. Generally, there a higher demand for LID in Montreal, with a slight difference compared to Toronto, and Vancouver has the lowest demand for LID. The mean HHI values for the C-base-00 and C-2096-02 scenarios range from 46.2 to 58.2 (a 26% increase) for Toronto, from 44.0 to 56.4 (a 28.2% increase) for Montreal, and from 21.3 to 25.1 (a 17.8% increase) for Vancouver.

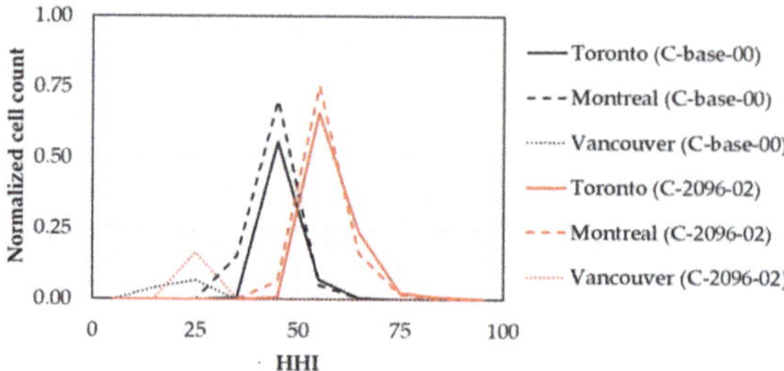

Figure 4. Normalized (to the maximum overall cell count of the three cities) histogram of HHI of comparing the current state (C-base-00) and the worst-case scenario (C-2096-02) among Toronto, Montreal, and Vancouver.

Toronto and Montreal present a similar pattern in response to climate change and urbanization. For example, in both cities, a significant proportion of the C-2096-02 histograms is higher than an HHI value of 50, which shows a higher demand for LID compared with the C-base-00 scenarios. In addition, there is a general change towards a higher frequency of HHI for the C-2096-02 from the C-base-00 scenario. In Vancouver, if an HHI value of 25 is used for comparison, the histogram does not show that the proportion of the C-2096-02 histogram for values greater than 25 has changed. Both the initial and final HHI values for Vancouver are much lower than both Toronto and Vancouver. The C-2096-02 demand is higher, as expected, for Vancouver, but with lower variance than the C-base-00 scenario.

The HHI values for all three cities are visualized in maps presented in Figures 5–7. As expected, all scenario results indicate that climate change and urbanization increase the HHI and, thus, the runoff generation potential in each city, leading to a higher demand for LID for source control. Climate change also significantly impacts HHI more than urbanization in Toronto and Montreal. This can be due to two reasons: first, Montreal is already highly urbanized and, thus, the room for future change in land cover is limited; and, second, while Toronto has a high share of pervious land cover, most of the natural soil has low permeability (clayey silt till or silty clay to silt till [63]), thus increasing the HHI further. Therefore, considering the current state of the land cover of these cities, urbanization has a lower effect than the change in the magnitude of rainfall intensity. Vancouver (results are shown in Figure 7) responded to the integrated climate change and urbanization similarly to the other two cities (an increase of HHI), but with lower values of HHI overall. The difference between the sole effect of urbanization and climate change in this city was not as significant as the other two cities. This is because Vancouver has a lower rainfall intensity compared to the other two cities, so the HHI value is about half the values for Toronto and Montreal, which decreases the effect of climate change. Note that Vancouver is more pervious compared to the other two cities, thus it has more room for urbanization (Figure 3).

Figure 8 compares the histogram of HHI for each city, for each urbanization scenario, for the four climate periods. The results show that, with time (from the base time to 2096), the HHI values increase (i.e., have a higher mean), i.e. the histograms shift to the right, which indicates higher runoff generation potential and, thus, an increase in demand for LID for source control. This shift is the lowest for Vancouver compared with the other two cities. There is an exception in the observed trend in Toronto for the 2066 scenario for which the mean HHI change compared with the previous scenario (2036) is minor due to the similarity of the rainfall intensity of these scenarios. Likewise, in Montreal, Scenario 2096 has slightly followed the overall pattern because the rainfall intensity in this scenario has not increased compared with the previous scenario (2066). In addition, with urbanization, the number of cells (i.e., locations within the cities) generating runoff is increasing as well. This is because pervious areas have been converted to impervious areas—and thus, are generating runoff in areas that did not under the base urbanization scenario (resulting in higher peaks of the histograms).

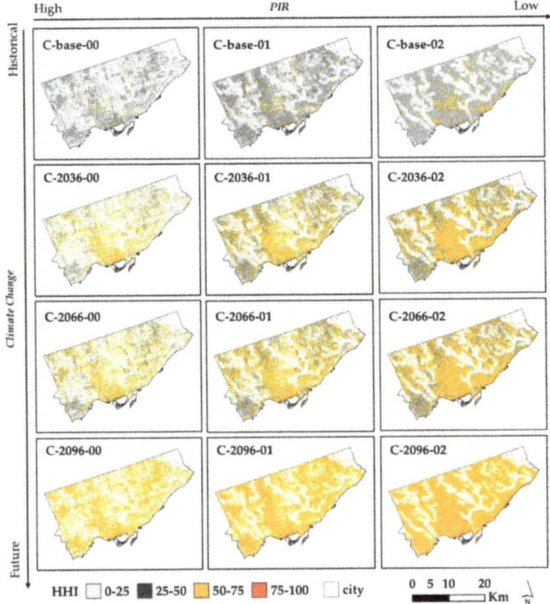

Figure 5. The HHI maps for Toronto under different climate (base year, 2036, 2066, and 2096) and land cover scenarios (base, Scenario 1, and Scenario 2).

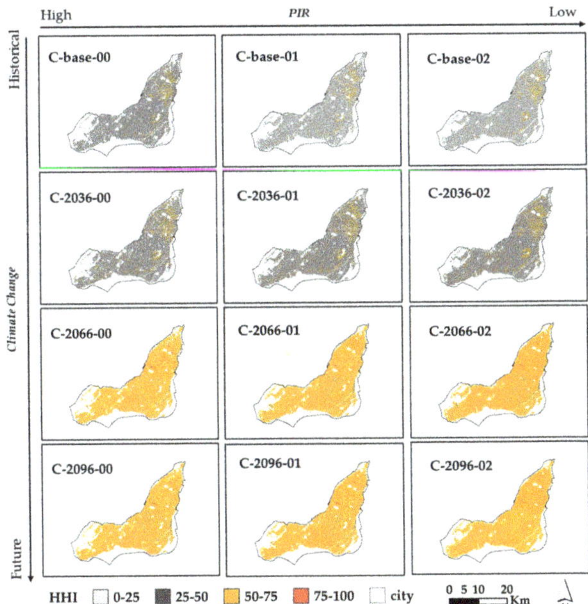

Figure 6. The HHI maps for Montreal under different climate (base year, 2036, 2066, and 2096) and land cover scenarios (base, Scenario 1, and Scenario 2).

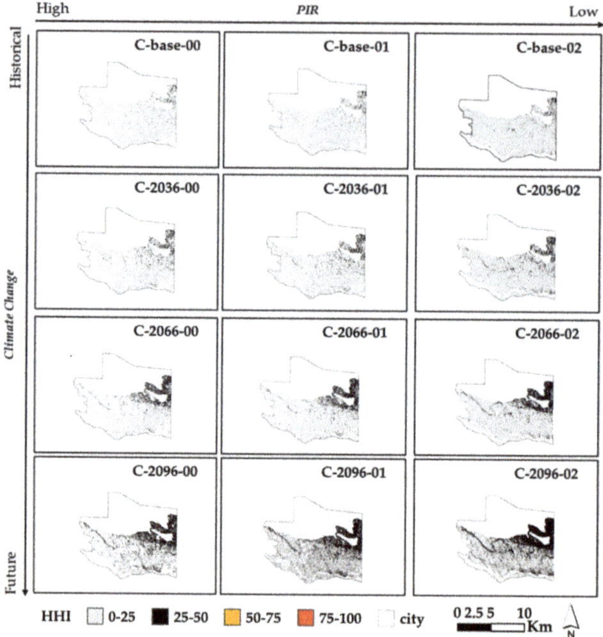

Figure 7. The HHI maps for Vancouver under different climate (base year, 2036, 2066, and 2096) and land cover scenarios (base, Scenario 1, and Scenario 2).

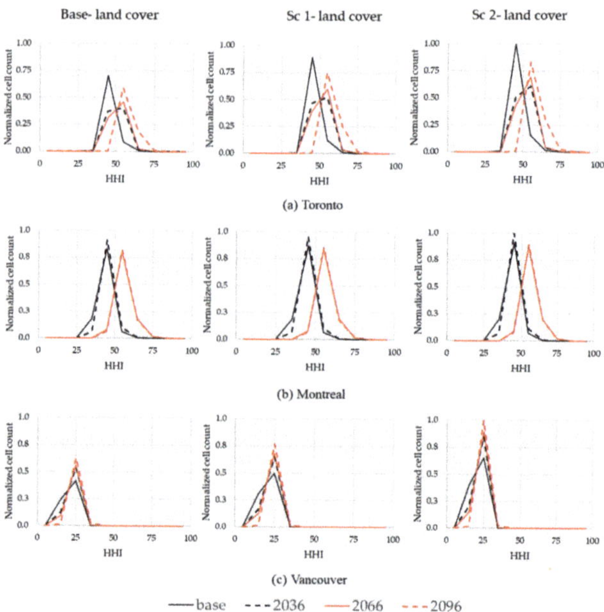

Figure 8. Normalized histograms of HHI (to the maximum HHI frequency of the cities) comparing the same urbanization scenarios (base, Scenario 1, and Scenario 2) and different Climate scenarios (base, 2036, 2066, and 2096) for: (**a**) Toronto; (**b**) Montreal; and (**c**) Vancouver.

Figure 9 shows a comparison of the climate change scenarios (i.e., rainfall intensity) for each city versus the different urbanization scenarios. This figure demonstrates that the change in climate has a relatively minor impact on the HHI frequency, as compared to the results shown in Figure 8. An increase in urbanization results in an increase in the frequency of HHI as well as an increase in the number of cells generating runoff; this is more evident in Toronto and Vancouver than in Montreal. While the increasing rainfall intensity for each city results in higher HHI values for each urbanization scenario, the change is relatively smaller than the effect illustrated in Figure 8.

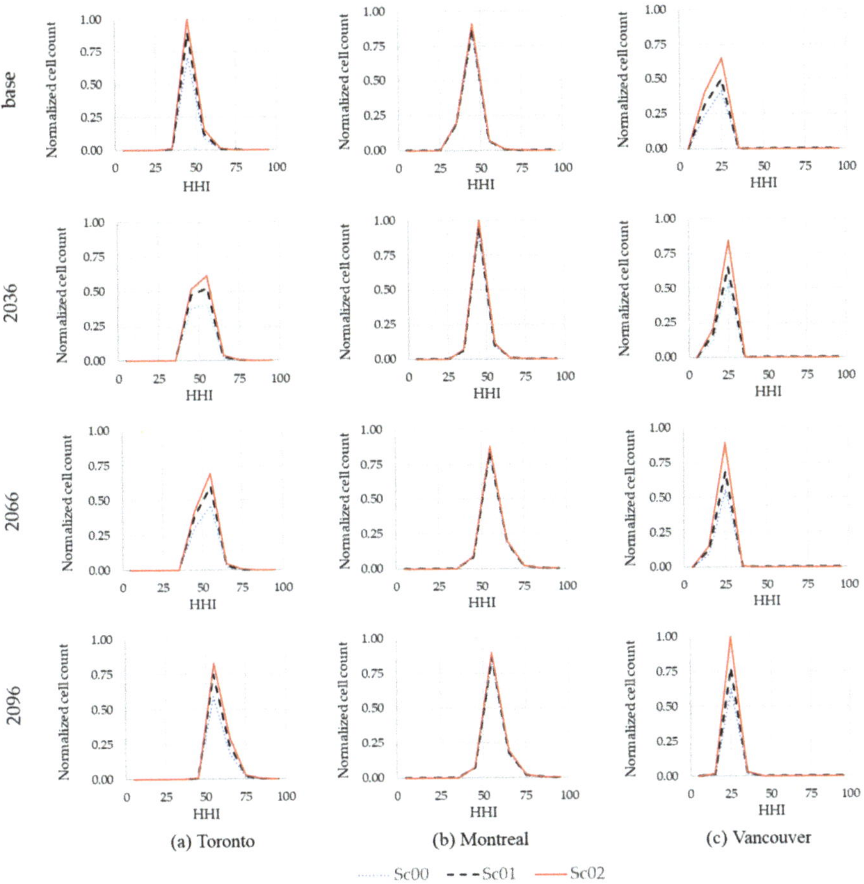

Figure 9. Normalized histograms of HHI (to the maximum HHI frequency of the cities) comparing the same climate scenarios (base, 2036, 2066, and 2096) and different urbanization scenarios (base, Scenario 1, and Scenario 2) for: (**a**) Toronto; (**b**) Montreal; and (**c**) Vancouver.

A comparison of all scenarios was conducted to determine the influence of urbanization and climate change on the magnitude of the area of sites that have a high demand for LID (referred to as high demand sites, HDS). To do this, the area of sites that have a value of HHI greater than a threshold value was calculated. The threshold was set at an HHI value of 50 for Toronto and Montreal and an HHI value of 20 for Vancouver (these values are the approximate mean HHI value for the C-base-00 scenario for each city). The area of HDS (AHDS) sites, the count of cells with an area of 25 m^2, was calculated for all scenarios and compared to AHDS of the baseline C-base-00 scenario. This comparison is shown in Figure 10.

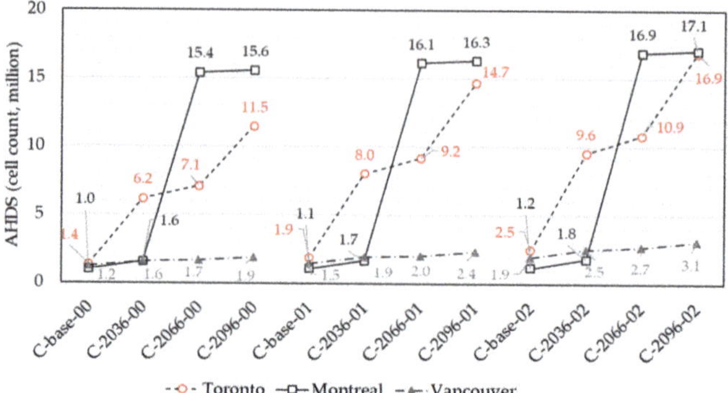

Figure 10. The total count of cells with HHI greater than the mean HHI for Toronto, Montreal, and Vancouver.

Overall, in Figure 10, all three cities show an increasing trend of AHDS under integrated climate change and urbanization as well as the sole effect of these scenarios. The overall pattern under the climate change only scenarios is similar to the corresponding rainfall patterns in Figure 2. For Toronto, urbanization seems to have a slight effect on AHDS, e.g., starting from 1.4 million cells for C-base-00 and reaching 2.5 million cells for C-base-02. In contrast, an increase in rainfall intensity due to climate change has a significant influence on increasing AHDS, e.g., the AHDS is 11.5 million cells for C-2096-00 and rises to 16.9 million cells for C-2096-02. Montreal presents similar behavior as Toronto; the AHDS of this city also has a notable increase with time for different climate scenarios. However, Vancouver's responses to climate change are similar to the impact of urbanization (from 1.2 million to 1.9 million cells in both cases) and are lower than in the other two cities. The reason is related to the low rainfall intensity of this city, which is a key difference between Vancouver and the other two cities.

To further investigate the sole effect of urbanization and climate change on each city, the normalized histogram of the worst case climate change scenario (C-2096-00) versus the worst case urbanization scenario (C-base-02) was compared, as presented in Figure 11. As concluded above, in response to the effects of climate change, Toronto and Montreal present an increase of HHI value with a shift of the histograms to the right, whereas this shift is lower in Vancouver. In terms of urbanization, it is evident that in all three cities the frequency of HHI has increased (i.e., higher peaks). As mentioned above, this increase is due to the increase in the number of cells that are generating runoff (non-zero runoff cells).

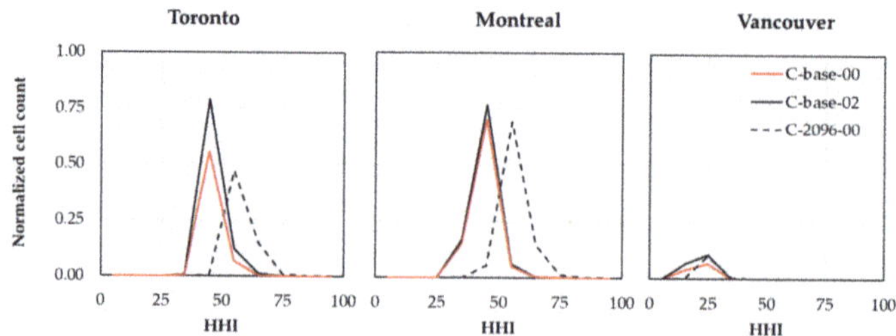

Figure 11. Normalized histograms of HHI (to the maximum count of three cities) of base scenario (C-base-00), the worst case of only the climate change scenario (C-2096-00) and the worst case of only urbanization (C-base-02) for the three cities (Toronto, Montreal, and Vancouver).

The trend of the mean HHI values for each city for all scenarios was also analyzed and is presented in Figure 12. The base scenario (C-base-00) shows that Toronto has a high HHI with a mean of 46.2.

The mean HHI increases to 57.7 (a 24.9% increase) due solely to climate change (C-2096-00) and slightly rises to 46.6 (a 0.9% increase) solely with urbanization (C-base-02). The integrated influence of climate change and urbanization raises HHI to 58.2 (C-2096-02) (a 26.0% increase). These results show that in Toronto the increase in mean HHI is due to climate change rather than urbanization. This is because the city has a high rainfall intensity and the natural soil has low permeability. The consequence of increasing the HHI is that the total area of sites that need LID is increasing (since more sites are generating runoff). In this respect, the AHDS of C-base-00 for this city is 1.4 million cells (0.2% of the city area) and reaches 11.5 million cells (8.4 times higher than the C-base-00) with the change of climate (C-2096-00). On the other hand, the AHDS growth is 78.6% by reaching 2.5 million cells with urbanization (C-base-02). The integrated effect of climate change and urbanization increases AHDS to 16.9 million, which is 12.3 times higher than AHDS of the current state.

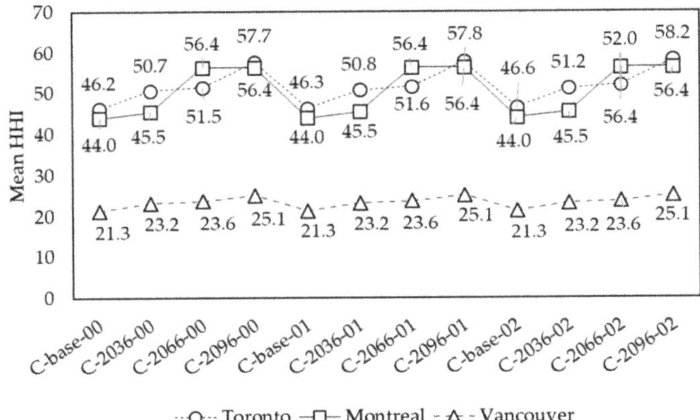

Figure 12. Comparing the trend of mean HHI of all twelve scenarios in Toronto, Montreal, and Vancouver.

Montreal has a mean HHI of 44.0 in the state of the base scenario. This value rises to 56.4 with only climate change (a 28.2% increase) and remains almost unchanged with only urbanization. The integrated effect of climate change and urbanization causes the increase of HHI value to the 56.4 (a 28.2% increase), which reveals the significant and dominant effect of climate change on HHI in Montreal. The highly urbanized land cover of Montreal leaves limited room for more urbanization; thus, the land cover of this city has a limited change in Scenarios 1 and 2. Thus, HHI in Montreal is more sensitive to climate change scenarios rather than urbanization. Regarding AHDS, in the base scenario, Montreal has AHDS of 1.0 million cells, which increases to 15.6 million cells with only climate change and 1.2 million cells with only urbanization. The AHDS for this city rises to 17.1 million cells in the integrated climate change and urbanization scenario.

The mean HHI value for Vancouver at the base scenario is 21.3, and this increases with climate change to 25.1 (a 17.8% increase) and remains unchanged with urbanization. The mean HHI rises to 25.1 (a 17.8%) for the integrated climate change and urbanization scenarios. The AHDS for this city at the current state is 1.2 million cells (0.9% of the city area), which increases to 1.9 million cells (1.6 times higher than the C-base-00) for only climate change scenarios and to 1.9 million cells (1.6 times higher than the C-base-00) for only urbanization scenario. The integrated climate change and urbanization scenario raises AHDS to 3.1 million cells, i.e., 2.5 times higher than the AHDS of the base scenario. The near-equal values of AHDS under the climate and urbanization scenarios indicate that the effect of both factors on HHI is similar. Although the mean HHI of urbanization scenarios remained the same, the increase of frequency of HHI has increased the overall AHDS.

4. Discussions

An overall comparison of the cities reveals the fact that the demand for LID (which is estimated using the HHI) in Toronto and Montreal is relatively similar and is about twice the demand for LID in Vancouver. This growth in demand is represented via AHDS: the AHDS increases from 2.5% to 16.8%, from the base scenario (C-base-00) to the worst-case scenario (C-2096-02), respectively, for all three cities, which indicates that all cities similarly respond to the integrated climate change and urbanization scenarios. The highest growth in AHDS is for Montreal (16.8 times higher than the C-base-00), followed by Toronto (12.3 times higher than the C-base-00), and last Vancouver (2.5 times higher than the C-base-00). The growth in LID demand is also shown in the increase of the mean HHI values, which increased by 26%, 28.2%, and 17.8% in Toronto, Montreal, and Vancouver, respectively. In addition, we showed that the demand for LID highly depends on the rainfall intensity rather than the overall mean annual rainfall. Vancouver, which has the highest mean annual precipitation among the selected cities, has the lowest demand for LID, which is due to its low rainfall intensity.

Generally, we showed that, if we retain the land cover as it is in these three cities, climate change will cause an increase in demand for LID since the runoff generation potential for each city will increase. With climate change only, urbanization only, and an integrated change of both, the three cities showed similar behavior: an increase in HHI, which indicates an increase in runoff generation potential, as has been reported in previous studies [64–70].

However, the standalone effects of climate change and urbanization were different for these cities. Toronto and Montreal were more sensitive to climate change compared with urbanization, and Vancouver was sensitive to both factors. Thus, depending on the rainfall intensity and the extent of urbanization in the study area the dominance of these two factors varies. Which factor is likely to have a greater influence on runoff generation has been reported in previous studies. Some studies found that urbanization likely exerts a greater impact on runoff than climate change [65,71–73], whereas other studies confirmed that climate change impact dominates the urbanization in runoff change [67,69,70]. Our finding is that the key factor is the rainfall intensity. Based on this, climate change dominates urbanization levels wherein the rainfall intensity is high (e.g., Toronto and Montreal) and adversely wherein the rainfall intensity is low (e.g., Vancouver). It is notable that, if the land cover of the study area is already impermeable (such as Toronto and Montreal), the dominance of climate change increases. In this context, the room for further urbanization is limited and climate change is the only factor that can alter the runoff.

In this study, we investigated the effect of poor urbanization by converting pervious land covers to impervious. By poor urbanization, we mean urban development by increasing impervious surfaces without either decreasing imperviousness in other locations or implementing LID. Therefore, for future research, we suggest the study of more city-scale cases such as areas with both high permeability and runoff intensity (which were not included in three case studies) to further investigate the dominance of urbanization and climate change.

In addition, in contrast to the poor urbanization scenarios, a study on smart urban growth [13] is recommended. From the hydrological point of view, smart growth consists of measures such as preventing urban sprawl in future development (e.g., poor urban planning such as designing wide streets) [13] as well as using integrated stormwater management strategies such as LID [74]. Smart growth is related to the early stages of urban planning [74] and strategies that need to be considered for future development. LID techniques, a component of smart growth, can be a solution for both pre- and post-development conditions, i.e., they can be retrofitted into highly urbanized areas or be designed in the initial planning stages of new developments. Cities similar to Montreal that are already highly urbanized have limited options for new developments wherein LID can be installed; however, LID can be retrofitted into existing areas. Retrofitting LID broadly across the city will help reduce the negative impacts of poor urbanization by increasing the overall permeability of the city. Cities such as Toronto, which are not as highly urbanized and have low permeability of the natural soil, have a high demand for LID. Thus, for this city, smart growth for future development as well as retrofitting existing areas with LID are both viable options. For Vancouver, which has a low demand for LID compared to the other two cities, the actual need for LID should be investigated in more detail using detailed design studies and hydrological modeling. However, similar to the other two cities, smart growth is suggested, since there is demand for LID and the impervious land cover in the city is dominant (61% of the city area).

Overall, the future urban growth (increase in the proportion of impervious surfaces) needs to be limited in order to reduce the runoff generation potential, and, hence, the risk of urban floods. Additionally, where possible, the current impervious land cover should be converted to natural, permeable surfaces. The increase in permeability (through the use of LID techniques) will help reduce runoff generation and offset the impacts of urbanization and climate change impacts.

5. Conclusions

In this study, we investigated the effect of climate change and urbanization on urban runoff generation potential using a geospatial framework for three Canadian cities (Toronto, Montreal, and Vancouver). The analysis was performed using a recently developed geospatial index (called HHI) using publicly available data. Four climate change scenarios (base, 2036, 2066, and 2096) and three urbanization scenarios (base land cover, Scenario 1, and Scenario 2) were developed. The HHI was estimated for the three cities for the 12 integrated climate change and urbanization scenarios.

In our study, we showed that climate change increases the mean HHI of the study area while urbanization raises the frequency of HHI. The extent of these changes varied based on the study area. Those with high rainfall intensity (Toronto and Montreal) showed a more significant increase in mean HHI compared with the study area with low rainfall intensity (Vancouver). In addition, the current impermeableness and room for further development within a city affected the extent of increasing the frequency of HHI.

The results agree with previous findings on the influence of climate change and urbanization on increasing the runoff and hence, the demand for LID. However, the magnitude of the effect of climate change and urbanization on the HHI varied in each city, which is also in agreement with results of previous studies. For example, in cities such as Toronto and Montreal with high rainfall intensity and low permeability, climate change is a dominant factor compared with urbanization. This is different from Vancouver that has low rainfall intensity climate change and urbanization level. In this city, HHI is sensitive to both factors. We also concluded that Toronto and Montreal have a higher demand for LID and this increases significantly with climate change and urbanization. Vancouver has a lower demand for LID and a lower rate of HHI change with the scenarios. The results of this study provide us with an insight into the contribution and the effect of climate and urbanization on the demand for LID.

These results, and the proposed methods, can be used in flood management, urban planning and for the sustainable development of cities. It is notable that, in addition to the rainfall and urbanization, HHI accounts for hydraulic conductivity, depth to the groundwater, and depth to the soil restrictive layer such as bedrock. In this study, we only investigated the effect of climate and urbanization on HHI and demand for LID, whereas modification of those parameters (hydraulic conductivity and depth to groundwater) in the future is possible and should also be studied.

Author Contributions: Conceptualization, S.K., U.T.K., and M.A.J.; Data curation, S.K.; Funding acquisition, U.T.K. and M.A.J.; Investigation, S.K.; Methodology, S.K., U.T.K., and M.A.J.; Resources, U.T.K. and M.A.J.; Software, S.K.; Supervision, U.T.K. and M.A.J.; Validation, S.K.; Visualization, S.K.; Writing—original draft, S.K.; and Writing—review and editing, U.T.K. and M.A.J. All authors have read and agreed to the published version of the manuscript.

Funding: This research was funded by the Natural Sciences and Engineering Research Council of Canada and York University.

Acknowledgments: The authors would like to thank the ESRI Inc. for granting the student version of ArcGIS. We would also like to thank York University Libraries for assisting with data collection. Lastly, we would like to thank three anonymous reviewers for their helpful suggestions which have led to an improved manuscript.

Conflicts of Interest: The authors declare no conflict of interest.

References

1. The United Nations Office for Disaster Risk Reduction (UNISDR). *The Human Costs of Weather Related Disasters 1995–2015*; The United Nations Office for Disaster Risk Reduction (UNISDR): Geneva, Switzerland, 2015; pp. 1–30.
2. Jha, A.K.; Bloch, R.; Lamond, J. *Cities and Flooding, A Guide to Integrated Urban Flood Risk Management for the 21st Century*; The World Bank: Washington, DC, USA, 2012; ISBN 978-0-8213-8866-2.
3. Aon Benfield. *2016 Annual Global Climate and Catastrophe Report*; Aon Benfield: London, UK, 2016; pp. 1–75.
4. Berndtsson, R.; Becker, P.; Persson, A.; Aspegren, H.; Haghighatafshar, S.; Jönsson, K.; Larsson, R.; Mobini, S.; Mottaghi, M.; Nilsson, J.; et al. Drivers of changing urban flood risk: A framework for action. *J. Environ. Manag.* **2019**, *240*, 47–56. [CrossRef]
5. Sarma, A.K.; Singh, V.P.; Kartha, S.A.; Bhattacharjya, R.K. *Urban Hydrology, Watershed Management and Socio-Economic Aspects*; Springer International Publishing: Basel, Switzerland, 2016; ISBN 9783319401942.
6. Ahiablame, L.; Shakya, R. Modeling flood reduction effects of low impact development at a watershed scale. *J. Environ. Manag.* **2016**, *171*, 81–91. [CrossRef] [PubMed]
7. Bailey, J.F.; Thomas, W.O.; Wetzel, K.L.; Ross, T.J. *Estimation of Flood-Frequency Characteristics and the Effects of Urbanization for Streams in the Philadelphia, Pennsylvania Area: U.S. Geological Survey Water-Resources Investigations Report 87-4194*; U.S. Geological Survey: Harrisburge, PA, USA, 1989; p. 76.
8. Brander, K.E.; Owen, K.E.; Potter, K.W. Modeled Impacts of Development Type on Runoff Volume and Infiltration Performance. *J. Am. Water Resour. Assoc.* **2004**, *40*, 961–969. [CrossRef]
9. Hamel, P.; Daly, E.; Fletcher, T.D. Source-Control Stormwater Management for Mitigating the Impacts of Urbanisation on Baseflow: A Review. *J. Hydrol.* **2013**, *485*, 201–211. [CrossRef]
10. Holman-Dodds, J.K.; Bradley, A.A.; Potter, K.W. Evaluation of Hydrologic Benefits of Infiltration Based Urban Storm Water Management. *J. Am. Water Resour. Assoc.* **2003**, *39*, 205–215. [CrossRef]
11. Bakhshipour, A.E.; Dittmer, U.; Haghighi, A.; Nowak, W. Hybrid green-blue-gray decentralized urban drainage systems design, a simulation-optimization framework. *J. Environ. Manag.* **2019**, *249*, 109364. [CrossRef] [PubMed]
12. Dietz, M.E. Low Impact Development Practices: A Review of Current Research and Recommendations for Future Directions. *Water Air Soil Pollut.* **2007**, *186*, 351–363. [CrossRef]
13. McCuen, R.H. Smart Growth: Hydrologic Perspective. *J. Prof. Issues Eng. Educ. Pract.* **2003**, *129*, 151–154. [CrossRef]
14. Coffman, L.S.; Goo, R.; Frederick, R. Low-Impact Development An Innovative Alternative Approach to Stormwater Management. In Proceedings of the 29th Annual Water Resources Planning and Management Conference (WRPMD), Tempe, AZ, USA, 6–9 June 1999; Volume 99, pp. 1–10.
15. Prince George's County Maryland. *Low-Impact Development Design Strategies An Integrated Design Approach Low-Impact Development: An Integrated Design Approach*; Department of Environmental Resources, Programs and Planning Division, Prince George's County: Prince George's County, MD, USA, 1999; pp. 1–150.
16. Heal, K.; Mclean, N.; D'arcy, B. SUDS and Sustainability. In Proceedings of the 26th Meeting of the Standing Conference on Stormwater Source Control, Copenhagen, Denmark, 21–26 August 2004; pp. 1–9.
17. Fletcher, T.D.; Shuster, W.; Hunt, W.F.; Ashley, R.; Butler, D.; Arthur, S.; Trowsdale, S.; Barraud, S.; Semadeni-Davies, A.; Bertrand-Krajewski, J.-L.; et al. SUDS, LID, BMPs, WSUD and more—The Evolution And Application Of Terminology Surrounding Urban Drainage. *Urban Water J.* **2015**, *12*, 525–542. [CrossRef]
18. Kaykhosravi, S.; Khan, U.; Jadidi, A. A Comprehensive Review of Low Impact Development Models for Research, Conceptual, Preliminary and Detailed Design Applications. *Water* **2018**, *10*, 1541. [CrossRef]
19. Ishaq, S.; Hewage, K.; Farooq, S.; Sadiq, R. State of provincial regulations and guidelines to promote low impact development (LID) alternatives across Canada: Content analysis and comparative assessment. *J. Environ. Manag.* **2019**, *235*, 389–402. [CrossRef] [PubMed]
20. Elliott, A.H.; Trowsdale, S.A. A Review Of Models For Low Impact Urban Stormwater Drainage. *Environ. Model. Softw.* **2007**, *22*, 394–405. [CrossRef]

21. Vogel, J.R.; Moore, T.L.; Coffman, R.R.; Rodie, S.N.; Hutchinson, S.L.; McDonough, K.R.; McLemore, A.J.; McMaine, J.T. Critical Review of Technical Questions Facing Low Impact Development and Green Infrastructure: A Perspective from the Great Plains. *Water Environ. Res.* **2015**, *87*, 849–862. [CrossRef]

22. Johns, C.; Shaheen, F.; Woodhouse, M. *Green Infrastructure and Stormwater Management in Toronto: Policy Context and Instruments*; Centre for Urban Research and Land Development: Toronto, ON, Canada, 2018; pp. 1–68.

23. Cheng, M.-S.; Coffman, L.S.; Clar, M.L. Low-Impact Development Hydrologic Analysis. *Urban Drain. Model.* **2001**, 659–681. [CrossRef]

24. Khan, U.T.; Valeo, C.; Chu, A.; van Duin, B. Bioretention Cell Efficacy In Cold Climates: Part 1—Hydrologic Performance. *Can. J. Civ. Eng.* **2012**, *39*, 1210–1221. [CrossRef]

25. Kaykhosravi, S.; Abogadil, K.; Khan, U.T.; Jadidi, M.A. The Low-Impact Development Demand Index: A New Approach to Identifying Locations for LID. *Water* **2019**, *11*, 2341. [CrossRef]

26. Kuller, M.; Bach, P.M.; Ramirez-Lovering, D.; Deletic, A. Framing water sensitive urban design as part of the urban form: A critical review of tools for best planning practice. *Environ. Model. Softw.* **2017**, *96*, 265–282. [CrossRef]

27. Kuller, M.; Bach, P.M.; Roberts, S.; Browne, D.; Deletic, A. A planning-support tool for spatial suitability assessment of green urban stormwater infrastructure. *Sci. Total Environ.* **2019**, *686*, 856–868. [CrossRef]

28. Liu, L.; Liu, Y.; Wang, X.; Yu, D.; Liu, K.; Huang, H.; Hu, G. Developing an effective 2-D urban flood inundation model for city emergency management based on cellular automata. *Nat. Hazards Earth Syst. Sci.* **2015**, *15*, 381–391. [CrossRef]

29. Esogbue, A.O.; Theologidu, M.; Guo, K. On the application of fuzzy sets theory to the optimal flood control problem arising in water resources systems. *Fuzzy Sets Syst.* **1992**, *48*, 155–172. [CrossRef]

30. Huang, C.; Hsu, N.; Liu, H.; Huang, Y. Optimization of Low Impact Development Layout Designs for Megacity Flood Mitigation. *J. Hydrol.* **2018**, *564*, 542–558. [CrossRef]

31. Haghighatafshar, S.; Yamanee-Nolin, M.; Klinting, A.; Roldin, M.; Gustafsson, L.G.; Aspegren, H.; Jönsson, K. Hydroeconomic optimization of mesoscale blue-green stormwater systems at the city level. *J. Hydrol.* **2019**, *578*, 124125. [CrossRef]

32. Eckart, K.B.C. Multiobjective Optimization of Low Impact Development Stormwater Controls under Climate Change Conditions. Master's Thesis, University of Windsor, Windsor, ON, Canada, 2015. Volume 562.

33. Ontario Ministry of Finance. *Ontario Population Projections, 2018–2046*; Ontario Ministry of Finance: Toronto, ON, Canada, 2019; pp. 3–111.

34. Government of Canada Climate Data for a Resilient Canada. Available online: https://climatedata.ca/ (accessed on 29 April 2020).

35. Institut De La Statistique Du Québec. *Forestière Française*; Institut De La Statistique Du Québec: Montreal, QC, Canada, 2019; pp. 1–18.

36. Metrovancouver Metro Vancouver Growth Projections—A Backgrounder. Available online: http://www.metrovancouver.org/services/regional-planning/PlanningPublications/OverviewofMetroVancouversMethodsinProjectingRegionalGrowth.pdf (accessed on 29 April 2020).

37. Government of Canada. Available online: https://www.canada.ca/en.html (accessed on 29 April 2020).

38. Government of Ontario. Available online: https://www.ontario.ca/page/government-ontario (accessed on 29 April 2020).

39. City of Toronto. Available online: https://www.toronto.ca/ (accessed on 29 April 2020).

40. City of Montreal National Hydrographic Network (NHN). Available online: http://donnees.ville.montreal.qc.ca/dataset/hydrographie/resource/8d563c07-02e0-4d2d-b1f9-733b132a6c03 (accessed on 29 April 2020).

41. Government of Quebec. Available online: http://www.environnement.gouv.qc.ca/index.asp (accessed on 29 April 2020).

42. Government of British Columbia. Available online: https://www2.gov.bc.ca/gov/content/home (accessed on 29 April 2020).

43. City of Vancouver Open Data Portal. Available online: https://opendata.vancouver.ca/pages/home/ (accessed on 29 April 2020).

44. Šraj, M.; Dirnbek, L.; Brilly, M. The Influence Of Effective Rainfall On Modeled Runoff Hydrograph. *J. Hydrol. Hydromech.* **2010**, *58*, 3–14. [CrossRef]

45. Bezak, N.; Šraj, M.; Rusjan, S.; Mikoš, M. Impact of the rainfall duration and temporal rainfall distribution defined using the Huff curves on the hydraulic flood modelling results. *Geosciences* **2018**, *8*, 1–15. [CrossRef]

46. Sikorska, A.E.; Viviroli, D.; Seibert, J. Effective precipitation duration for runoff peaks based on catchment modeling. *J. Hydrol.* **2018**, *556*, 510–522. [CrossRef]

47. City of Toronto. *Wet Weather Flow Management Guidelines*; City of Toronto: Toronto, ON, Canada, 2006; pp. 1–117.

48. The City of Edmonton. *Low Impact Development Best Management Practices Design Guide Edition 1.1*; The City of Edmonton: Edmonton, AB, Canada, 2014; pp. 1–269.

49. City of Saskatoon. *Low Impact Development: Design Guide for Saskatoon Prepared by City of Saskatoon November 2016 Preface*; City of Saskatoon: Saskatoon, SK, Canada, 2016; pp. 1–64.

50. Toronto and Region Conservation for The Living City; Credit Valley Conservation. *Low Impact Development Stormwater Management Planning and Design Guide Version 1.0*; Toronto and Region Conservation Authority: Toronto, ON, Canada, 2010; pp. 1–300.

51. Institute for Catastrophic Loss Reduction; Western University Computerized Tool for the Development of Intensity-Duration-Frequency Curves under Climate Change—Version 4.0, IDF_CC Tool 4.0. Available online: https://www.idf-cc-uwo.ca/home.aspx (accessed on 29 April 2020).

52. IDF_CC Tool 4.0, Computerized Tool for the Development of Intensity-Duration-Frequency Curves Under a Changing Climate, Technical Manual v.3. Available online: https://www.eng.uwo.ca/research/iclr/fids/publications/products/103.pdf. (accessed on 29 April 2020).

53. IDF_CC Tool 4.0, Computerized Tool for the Development of Intensity-Duration-Frequency Curves Under a Changing Climate, User's Manual v.3. Available online: https://www.eng.uwo.ca/research/iclr/fids/publications/products/104.pdf. (accessed on 29 April 2020).

54. van Vuuren, D.P.; Edmonds, J.; Kainuma, M.; Riahi, K.; Thomson, A.; Hibbard, K.; Hurtt, G.C.; Kram, T.; Krey, V.; Lamarque, J.F.; et al. The representative concentration pathways: An overview. *Clim. Chang.* **2011**, *109*, 5–31. [CrossRef]

55. Coulibaly, P.; Burn, D.H.; Switzman, H.; Henderson, J.; Fausto, E. *A Comparison of Future IDF Curves for Southern Ontario Addendum—IDF Statistics, Curves and Equations*; Toronto and Region Conservation Authority (TRCA); Essex Region Conservation Authority (ERCA): Toronto, Canada, 2016; pp. 1–244.

56. *Climate Change (2050) Adjusted IDF Curves: Metro Vancouver Climate Stations*; BGS Engineering Inc.: Vancouver, BC, Canada, 2009; pp. 1–36.

57. Jadidi, A.; Mostafavi, M.; Bédard, Y.; Shahriari, K. Spatial Representation of Coastal Risk: A Fuzzy Approach to Deal with Uncertainty. *ISPRS Int. J. Geo-Inf.* **2014**, *3*, 1077–1100. [CrossRef]

58. Li, X.; Gong, P. Urban growth models: Progress and perspective. *Sci. Bull.* **2016**, *61*, 1637–1650. [CrossRef]

59. Metro Vancouver Metro Vancouver Open Data. Available online: http://www.metrovancouver.org/data (accessed on 29 April 2020).

60. Chabaeva, A.; Civco, D.; Prisloe, S. Development of a population density and land use based regression model to calculate the amount of imperviousness. In Proceedings of the 2004 ASPRS Annual Convention Denver, Denver, CO, USA, 23–28 May 2004; p. 9.

61. Huang, H.; Cheng, S.; Wen, J.; Lee, J. Effect of growing watershed imperviousness on hydrograph parameters and peak discharge. *Hydrol. Process.* **2007**, *22*, 2075–2085. [CrossRef]

62. Ontario Ministry of Finance. *Ontario Population Projections, 2018–2046: Based on the 2016 Census, for Ontario and Its 49 Census Divisions*; Ontario Ministry of Finance: Toronto, ON, Canada, 2019; pp. 1–109.

63. Ministry of Northern Development and Mines. Province of Ontario's Ministry of Northern Development and Mines (MNDM). Available online: http://www.geologyontario.mndm.gov.on.ca/mndmfiles/pub/data/imaging/P2204/p2204.pdf (accessed on 29 April 2020).

64. Yin, J.; He, F.; Jiu Xiong, Y.; Yu Qiu, G. Effects of land use/land cover and climate changes on surface runoff in a semi-humid and semi-arid transition zone in northwest China. *Hydrol. Earth Syst. Sci.* **2017**, *21*, 183–196. [CrossRef]

65. Zhou, Q.; Leng, G.; Su, J.; Ren, Y. Comparison of urbanization and climate change impacts on urban flood volumes: Importance of urban planning and drainage adaptation. *Sci. Total Environ.* **2019**, *658*, 24–33. [CrossRef] [PubMed]

66. Zhang, Y.; Xia, J.; Yu, J.; Randall, M.; Zhang, Y.; Zhao, T.; Pan, X.; Zhai, X.; Shao, Q. Simulation and assessment of urbanization impacts on runoff metrics: Insights from landuse changes. *J. Hydrol.* **2018**, *560*, 247–258. [CrossRef]

67. Zhou, Y.; Lai, C.; Wang, Z.; Chen, X.; Zeng, Z.; Chen, J.; Bai, X. Quantitative evaluation of the impact of climate change and human activity on runoff change in the Dongjiang River Basin, China. *Water (Switzerland)* **2018**, *10*, 571. [CrossRef]

68. Al-Zahrani, M.A. Assessing the impacts of rainfall intensity and urbanization on storm runoff in an arid catchment. *Arab. J. Geosci.* **2018**, *11*, 208. [CrossRef]

69. Yang, W.; Long, D.; Bai, P. Impacts of future land cover and climate changes on runoff in the mostly afforested river basin in North China. *J. Hydrol.* **2019**, *570*, 201–219. [CrossRef]

70. Arnone, E.; Pumo, D.; Francipane, A.; La Loggia, G.; Noto, L.V. The role of urban growth, climate change, and their interplay in altering runoff extremes. *Hydrol. Process.* **2018**, *32*, 1755–1770. [CrossRef]

71. Wang, M.; Zhang, D.Q.; Su, J.; Trzcinski, A.P.; Dong, J.W.; Tan, S.K. Future Scenarios Modeling of Urban Stormwater Management Response to Impacts of Climate Change and Urbanization. *Clean Soil Air Water* **2017**, *45*, 1700111. [CrossRef]

72. Wang, M.; Zhang, D.; Adhityan, A.; Ng, W.J.; Dong, J.; Tan, S.K. Assessing cost-effectiveness of bioretention on stormwater in response to climate change and urbanization for future scenarios. *J. Hydrol.* **2016**, *543*, 423–432. [CrossRef]

73. Gao, C.; Ruan, T. The influence of climate change and human activities on runoff in the middle reaches of the Huaihe River Basin, China. *J. Geogr. Sci.* **2018**, *28*, 79–92. [CrossRef]

74. Ahiablame, L.M.; Engel, B.A.; Chaubey, I. Effectiveness of low impact development practices: Literature review and suggestions for future research. *Water Air Soil Pollut.* **2012**, *223*, 4253–4273. [CrossRef]

Article

Trends and Non-Stationarity in Groundwater Level Changes in Rapidly Developing Indian Cities

Aadhityaa Mohanavelu [1], K. S. Kasiviswanathan [2,*], S. Mohanasundaram [3], Idhayachandhiran Ilampooranan [2], Jianxun He [4], Santosh M. Pingale [5], B.-S. Soundharajan [1] and M. M. Diwan Mohaideen [2]

[1] Department of Civil Engineering, Amrita School of Engineering, Amrita Vishwa Vidyapeetham, Coimbatore 641112, India; aadhityaa65@gmail.com (A.M.); b_soundharajan@cb.amrita.edu (B.-S.S.)
[2] Department of Water Resources Development and Management, Indian Institute of Technology Roorkee, Roorkee 247667, India; idhaya@wr.iitr.ac.in (I.I.); diwan1990@gmail.com (M.M.D.M.)
[3] Water Engineering and Management, School of Engineering and Technology, Asian Institute of Technology, P.O. Box 4 Klong Luang, Pathum Thani 12120, Thailand; mohanasundaram@ait.ac.th
[4] Department of Civil Engineering, Schulich School of Engineering, University of Calgary, Calgary, AB T2N 1N4, Canada; jianhe@ucalgary.ca
[5] Hydrological Investigations Division, National Institute of Hydrology, Roorkee 247667, India; pingalesm@gmail.com
* Correspondence: k.kasiviswanathan@wr.iitr.ac.in

Received: 29 September 2020; Accepted: 11 November 2020; Published: 16 November 2020

Abstract: In most of the Indian cities, around half of the urban water requirement is fulfilled by groundwater. Recently, seasonal urban droughts have been frequently witnessed globally, which adds more stress to groundwater systems. Excessive pumping and increasing demands in several Indian cities impose a high risk of running out of groundwater storage, which could potentially affect millions of lives in the future. In this paper, groundwater level changes have been comprehensively assessed for seven densely populated and rapidly growing secondary cities across India. Several statistical analyses were performed to detect the trends and non-stationarity in the groundwater level (GWL). Also, the influence of rainfall and land use/land cover changes (LULC) on the GWL was explored. The results suggest that overall, the groundwater level was found to vary between ±10 cm/year in the majority of the wells. Further, the non-stationarity analysis revealed a high impact of rainfall and LULC due to climate variability and anthropogenic activities respectively on the GWL change dynamics. Statistical correlation analysis showed evidence supporting that climate variability could potentially be a major component affecting the rainfall and groundwater recharge relationship. Additionally, from the LULC analysis, a decrease in the green cover area (R = 0.93) was found to have a higher correlation with decreasing groundwater level than that of urban area growth across seven rapidly developing cities.

Keywords: groundwater level; trends; non-stationarity; climate variability; land use/land cover change; developing cities

---

1. Introduction

Groundwater, one of the Earth's largest available freshwater resources, is being extracted at a rate of ~982 km^3/year [1,2]. Globally, groundwater meets about half of domestic water needs and 38% of irrigation water demand for sustaining food security [3,4]. The increasing global population is a major concern since presently, about 80% of the world's population is at an incident threat to water security [5]. Unlike developed countries, developing countries like India are at high risk of water scarcity due to the increasing population, lack of adequate regulatory policies and investments in water technology, intense anthropogenic activities, and growing urban settlements [5]. In 2050, globally, more than two-thirds of the population is expected to live in urban areas, of which urban India alone will be the home to 14% of the world's population [6,7]. In the next thirty years, it is anticipated that India should come up with principal changes in urban life as half of India's population would undergo tremendous changes in the landscape and socio-economic structure [7]. Thus, ensuring water security is the need of the hour to decide the course of a country's growth and future [8].

Globally, the increase in urban groundwater use is driven by the reduction in surface water potential, higher water-supply costs from public supply schemes, easy access to high-yielding aquifers, a relatively low cost involved in well construction and maintenance, private ownership, and high reliability (availability of water even during dry seasons) [9]. Presently, it is estimated that groundwater depletion has increased in 40% of the major cities around the world [10]. In Brazil, a major increase in the privately owned wells was reported during the period 1995–2010, as a response to mitigate the inadequacy in the water-supply during the extended drought crisis [11]. In Sub-Saharan Africa, groundwater is the fast-growing source of urban water supply to meet the proliferating demand, despite the higher costs involved in well drilling [11]. Even in some developed countries in Europe, about 40% of the urban supply comes from groundwater [12]. It is important to note that limitations in both financial and geographical conditions are the important factors in determining urban water scarcity. Cities that lack coordination among different stakeholders and financial limitations (e.g., Dhaka in Bangladesh and Kampala in Uganda, etc.) prevents the development of a large-scale water supply project [12,13]. Under such limitations, groundwater seems to be a practically viable source, and people in such cities prefer drilling water-wells to satisfy partial or all of their water requirements. Globally, this trend is becoming common in developing cities due to high population growth, rapid urbanization, increasing per-capita usage of water, high ambient temperatures, and reduced intake from surface sources (due to increased risk of pollution) [13].

Recent studies suggest that groundwater levels are declining in several parts of northern India, especially in the regions of high population densities [14–17]. On the contrary, increasing groundwater levels have been reported in Southern India in recent studies, although well failures and groundwater stress are commonly reported in these regions [16,18,19]. Further, a high irrigation demand to boost the food supply has already led to the over-exploitation of groundwater, especially in high-intense agriculture states such as Punjab and Haryana [20,21]. The dependence of groundwater for public and private water supply is also becoming indispensable in developing cities across India, which would further add stress on groundwater systems [22,23].

According to the report of the National Institute for Transforming India (NITI) Aayog's Composite Water Management Index (CWMI), twenty-one Indian cities, including major cities like Delhi, Bengaluru, Chennai, and Hyderabad, are presently facing an acute water crisis, and about 40% of India's population possibly would have no access to drinking water by 2030 [24]. The recent water crisis during the 2019 summer in the Indian city of Chennai has highlighted the imminent water crisis. Chennai's groundwater storage fell rapidly due to limited recharge from the monsoon, and long-term over-exploitation, which led to drying up of wells, and the city has ferried water from nearby areas through tankers and rail wagons to suffice the seasonal demand [25]. This is a clear warning to most of the other Indian cities, which are highly dependent on groundwater (unreliable public water supply systems often fail due to frequent urban drought) for meeting the domestic water demands. In most of the urban areas, water supplied through the public water supply system is inadequate due to water scarcity or lack of water supply infrastructure [26]. Thus, the benchmark supply level of 135 L per capita per day (lpcd), prescribed by the Indian Ministry of Housing and Urban Development, could not be met [26]. Therefore, the groundwater through household or private bore wells has often been used to offset the demand–supply gap. The limited water availability imposes the risk of over-exploitation of groundwater to meet the domestic and industrial water demands, along with real-estate growth, and thus causes a huge threat to the groundwater reserve [27,28]. While climate change and anthropogenic activities affect water availability [8], the high dependency of groundwater in larger urban centers of

India has created an acute supply–demand deficit [29]. Recent studies also suggest that the frequency of droughts over the massive urban environments is likely to increase throughout the Indian cities in the near future [30,31].

The migration to cities from rural areas is exorbitant and likely to increase in the future. Thus, the projected rapid urban growth is concentrated particularly in secondary rapidly developing cities. This would consequently lead to unprecedented challenges and uncertainties in meeting water needs in the future. While there is a little scope for improving the water infrastructure in developed cities conceivably due to feasibility and implementation issues, proper planning and management of water systems are crucial in secondary developing cities.

In the Indian context, no study reported the trends in groundwater level (GWL) changes relating to non-stationarity neither for developed cities nor secondary cities. However, in other parts of the world, the spatio-temporal changes and non-stationarity in groundwater levels have been analyzed using several statistical methods [32–35]. The major findings from these studies are (i) majority of the wells exhibited the non-stationarity with significant trends in GWL changes, and (ii) precipitation and recharge of groundwater levels are affected by climate variability (both on an Annual and Decadal scale), which has more impact than the increasing temporal patterns in pumping [32–35].

While previous studies focused on assessing the GWL trends influenced by the land use/land cover (LULC) changes at multiple spatial scales [36,37], this paper focused on a comprehensive analysis of GWL change dynamics considering the climate variability (rainfall) as well as LULC change (anthropogenic activity), especially in rapidly developing cities, and discussed the potential driving forces responsible for non-stationarity. For this endeavor, seven densely populated and rapidly growing secondary cities that share similar socio-economic developmental goals were selected from different geographic regions across India for assessing the condition of groundwater systems. Various statistical analyses, widely employed in the literature, such as Theil-Sen estimator [38], modified Mann–Kendall test [39], Augmented Dicky Fuller test [40], Phillip-Perron test [41], and Kwiatkowski-Phillips-Schmidt-Shin test [42], were used to determine the trends and non-stationarity in the groundwater level (GWL) changes [32,33,35,43,44]. The specific objectives of the paper are to (i) provide an overall outlook of trends in GWL change dynamics, (ii) compare the GWL changes with the Total Water Storage (TWS) anomalies, (iii) determine the non-stationarity in the GWL changes, and (iv) assess the contribution of climate variability (rainfall) and LULC change to the non-stationarity.

2. Materials and Methods

2.1. Study Area and Data Collection

Seven cities (Table 1) were selected based on the following criteria: (i) population ~1 Million (2011 Census) [45], (ii) decadal population growth rate greater than ~15% (2001 to 2011 Census) [45], (iii) high Gross Domestic Product (GDP) and economic growth rate [46], (iv) population density, and (v) shortlisted as a smart city under the urban renewal program, National Smart City Mission, by the government of India [47]. The groundwater level data for the years 1996–2018 has been collected for the selected seven cities from the Central Groundwater Board (CGWB) through the India Water Resource Information system [48]. The CGWB monitors seasonal groundwater levels quarterly (i.e., January—Post-monsoon Rabi, May—Pre-monsoon, August—Monsoon, and November—Post-monsoon) [49]. Grubbs test was used to remove the outliers in the dataset. The Grubbs test statistics are defined as $G = \frac{max|Y_i - \overline{Y}|}{s}$, where Y_i is the ith data point and $\overline{Y}$ and s denote the sample mean and standard deviation, respectively. The wells located in the selected cities having at least 18 years of observed data out of a total of 23 years (~80% observations) were selected for the analysis (refer to Figure 1 for the map of the study area and the wells selected for the analyses) [50]. The monthly Terrestrial Water Storage anomalies were obtained from Gravity Recovery and Climate Experiment (GRACE) satellite observations for the months of January, May, August, and November within the study period (2002–2017) to analyze the total water storage in the selected cities [51]. Daily gridded rainfall data (0.25° × 0.25°) obtained from the Indian Meteorological Department (IMD) were used to estimate the monthly rainfall values for the selected cities [52]. Moderate Resolution Imaging Spectroradiometer (MODIS) classified LULC maps (MCD12Q1v006: LC_type 4 layer) obtained for the years 2001 and 2018 were used to perform the LULC analysis [53].

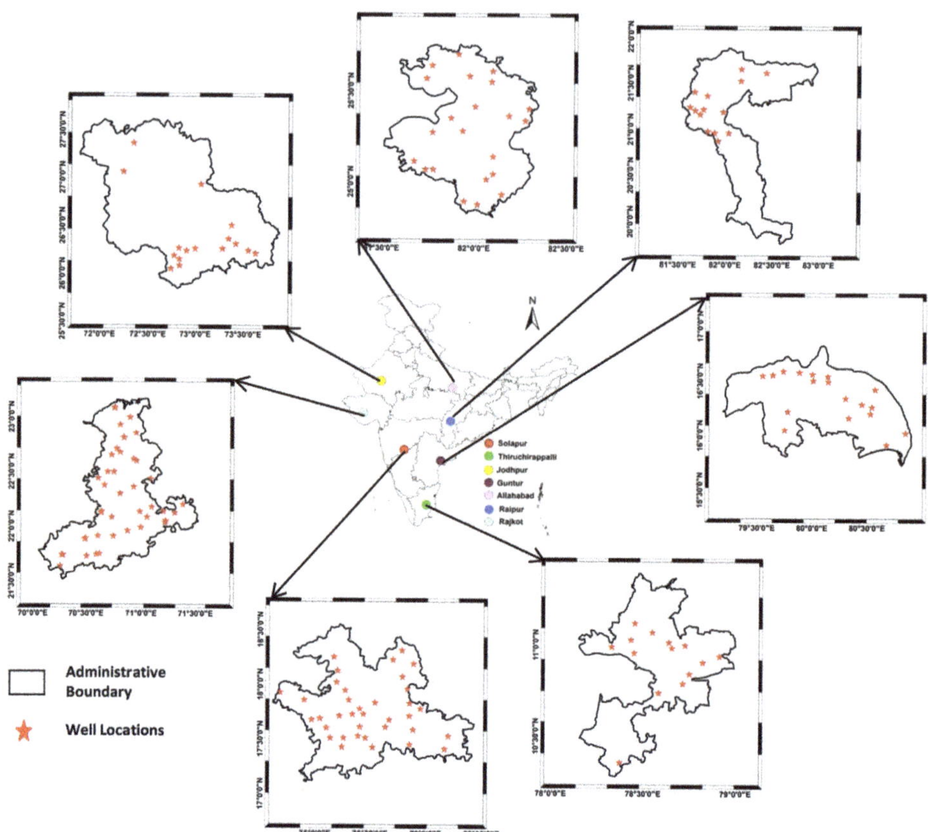

Figure 1. Study area map with the well locations in seven cities.

Table 1. Population and observation well details

City	State	Population (Census 2011)	Population Growth Rate, % (2001–2011 Census)	Population Density (No. of People Per sq. km)	Number of Observatory Wells Studied
Allahabad	Uttar Pradesh	1,112,544	21	14,000	31
Guntur	Andhra Pradesh	743,354	45	430	21
Jodhpur	Rajasthan	1,033,918	28	4900	19
Raipur	Chhattisgarh	1,010,087	35	4500	26
Solapur	Maharashtra	951,558	12	5300	49
Tiruchirappalli	Tamil Nadu	916,674	13	5500	17
Rajkot	Gujarat	1,286,678	20	110	54

2.2. Trend Analysis

The Theil-Sen estimator, a widely used method in detecting trends in a hydrological time series, was employed to determine the nonparametric linear GWL trend (in cm year^{-1}) [38,54], and the significance of the determined trend was estimated by applying the modified Mann–Kendall (MMK) test [39,55]. The Theil-Sen estimator determines the slope between all possible data pairs as, $Q_i = \frac{y_j - y_k}{j - k}$, where Q_i is the slope between data points y_j and y_k at time j and k, such that $j > k$ and $i = 1, 2 \ldots n$. For n values of the time series of y number of slopes, there is $N = \frac{n(n-1)}{2}$ values of Q_i. The Theil-Sen estimator determines the median slope (Q) as, $Q = Q_{\left(\frac{n+1}{2}\right)}$ if N is odd and $Q = \frac{1}{2}\left(Q_{\left[\frac{N}{2}\right]} + Q_{\left[\frac{N+1}{2}\right]}\right)$ if N is even [38]. The MMK test uses a modified variance to reduce the influence of autocorrelation on the Mann–Kendall test results [56,57]. The test statistics of MMK are the same as the Mann–Kendall test and are defined as $S = \sum_{i=1}^{n-1}\sum_{j=k+1}^{n} Sign(x_j - x_k)$, where S is the test statistics, x_j and x_k are the data points observed at time j and k respectively ($j > k$), and n is the length of the dataset [39,56]. The trends in the GWL were estimated in terms of changes in the depth to groundwater table below ground level in centimeters per year. Although several methods exist to recover the GWL from Satellite products (e.g., GRACE [51]), these methods provide only gross estimates with higher uncertainty, bias, and errors [58,59]. To overcome the above issues, GWL observations from monitoring wells by CGWB were used. Though the trend analysis was focused on estimating the temporal trends at the point level (individual wells), the selected wells are uniformly distributed across the city (Figure 1), with each well representing a block (or part) of a city, which subsequently helps to understand the spatial trends as well.

2.3. TWS Anomalies and Groundwater Levels

To understand the extent of changes in the overall water storage (includes surface water, soil moisture, groundwater, etc.), the Terrestrial Water Storage (TWS) anomalies were analyzed. The TWS anomalies data were obtained from the Gravity Recovery Climate Experiment (GRACE) interpolated at 1° × 1° grid cells for the selected seven cities for the period 2002 to 2017 [51]. The trends and non-stationarity in the TWS changes were determined, and the results of TWS were correlated with GWL changes in the selected cities.

2.4. Non-Stationarity Analysis

Non-stationarity analyses were performed to assess the impacts of climate change caused by low-frequency climate variability and LULC changes on GWL changes [60]. The most frequently used Augmented Dickey-Fuller (ADF), Phillip-Perron (PP), and Kwiatkowski-Phillips-Schmidt-Shin (KPSS) tests were applied to estimate the non-stationarity in GWL changes [40–42].

The ADF test assesses the null hypothesis of a unit root using the model: $y_t = y_t - 1 + \beta_1\Delta y_t - 1 + \beta_2\Delta y_t - 2 + \ldots + \beta_p\Delta y_t - p + \varepsilon_t$ against the alternative model: $y_t = \phi y_t - 1 + \beta_1\Delta y_t - 1 + \beta_2\Delta y_t - 2 + \ldots + \beta_p\Delta y_t - p + \varepsilon_t$, where y_t is the time series with time t, Δ is the differencing operator, β is constant, p is the number of lagged difference terms, and ε_t is a mean zero innovation process [40,61]. The alternate hypothesis ($\phi < 1$) in the ADF test is generally stationary or trend stationary [61].

Similar to the ADF test, the PP test assesses the null hypothesis of a unit root in a univariate time series; however, it makes a nonparametric correction to t-statistics, which makes it more robust with unspecified autocorrelation [62,63]. The PP test assesses the null hypothesis using the model: $y_t = y_{t-1} + e(t)$, against the alternative model: $y_t = a\, y_{t-1} + e(t)$, where y_t is the time series with time t, $e(t)$ is the innovations process and a is the auto regression coefficient, such that $a < 1$ [41,63].

The KPSS test, in contrast to ADF and PP tests (which tests for non-stationary as a null hypothesis), assesses the null hypothesis that a univariate time series is trend stationary against the alternate hypothesis of non-stationarity [42]. The KPSS test assesses the null hypothesis of trend stationarity using the models: $y_t = c_t + \delta_t + u_{1t}$ and $c_t = c_t - 1 + u_{2t}$ against the alternate hypothesis of non-stationarity, where c_t is the random walk term, δ is the trend coefficient, u_{1t} is a stationary process, and u_{2t} is an independent and identically distributed process with mean 0 and variance σ^2 [64]. The null hypothesis $\sigma^2 = 0$ implies that the random walk term (c_t) is constant, and the alternative hypothesis $\sigma^2 > 0$, which introduces the unit root in c_t [42,64].

Although the power of non-stationarity tests varies, all the non-stationarity tests are known to have low power (power is determined by auto-regression parameter ϕ_1 in case of ADF and PP tests) when the length of the time series is short [65]. However, the KPSS test is an exception and performs well in shorter time series, which made it suitable for our analysis (although the KPSS test has a high rate of Type I Errors (rejection of null hypothesis)) [65,66]. As the availability of the observed GWL data is very limited, the non-stationarity in the GWL changes was confirmed if the majority of the three tests (ADF, PP, and KPSS) suggest non-stationarity. Thus, the bias due to data length was minimized.

2.5. Correlation Analysis: Rainfall and Groundwater Levels

The impacts of climate variability on GWL changes were grossly analyzed by correlating rainfall with GWL changes in the wells having non-stationarity [67]. The relationship between rainfall–groundwater recharge is complex; however, few studies show a high correlation between them [68–70]. Since the groundwater system's recharge characteristics are complex, there is always a time delay between rainfall and the water to reach the groundwater table [68,71]. The cross-correlation was carried out for two conditions such as: (i) no lag and (ii) 3-month lag (or 1 season lag) with the groundwater level data to account for the delay in the recharge of groundwater after the rainfall. For this, the gridded rainfall data and GWL changes from the wells exhibiting non-stationarity behavior and falling within the grid were used. It may be noted that about 94% of the wells used in the analysis are dug wells (lies predominantly in the unconfined aquifers) with depth generally varying between 10 and 30 m, and therefore directly relating the monthly rainfall to GWL changes is justified (not accounting for the recharge from the confined aquifer) [49].

2.6. LULC Change Analysis

The impacts of LULC change on GWL were also quantified. The Moderate Resolution Imaging Spectroradiometer (MODIS) classified LULC maps (MCD12Q1v006: LC_type 4 layer) for the years 2001 and 2018 were used to analyze the LULC changes in the respective cities. This MODIS-MCD12Q1v006 data product for global land cover is available at 500 m resolution [53]. A 15 km radial buffer around the center point of the cities was created since the overlay analysis revealed that the urban expansion between 2001 and 2018 at these selected cities was very much contained within this buffer zone. This 15 km buffer zone shapefile was used to clip out the LULC maps of the respective cities. Google Earth Engine (GEE) algorithms were used to analyze the entire spatial data, starting from accessing the data to the creation of the LULC change map. The impact of change in LULC components on GWLs was assessed through performing correlation analysis between the percentage of wells having significant increasing/decreasing GWL trends (>2 cm/year) and LULC change between 2001 and 2018 at the city level. The statistical analyses were also performed for the wells located outside the urban boundary as these cities have high scope for expansion in terms of infrastructural growth and socio-economic activities.

3. Results and Discussion

3.1. Trends in Groundwater Level Changes

Trend analysis for GWL changes revealed that the GWL majorly varied between ±~10 cm/year (Figure 2) across the selected cities. Spatially, GWL trends (Figure 3) were found to be decreasing in the majority of the wells, except in Rajkot and Guntur. In general, consistent high decreasing and low increasing GWL trend patterns were observed in January relatively, compared to other months of observation (Figure 2). A high increasing GWL trend was found in May (Figures 2–4), though May is considered to have peak summer with high evapotranspiration losses and very few rain spells in India [72,73]. This anomaly, however, can be attributed to changes in the characteristics and spatial-temporal variability of rainfall events in India [74,75]. Very similar GWL trends across the seasonal data collected in January, May, August, and November (Figure 2) were observed, which suggests a less interdependence between seasonality and GWL trends. However, the lower tail of the box plot (Figure 2) indicates a sharp increase in GWL, and this behavior could be attributed to groundwater recharge from the monsoon rainfall [15,16]. Similarly, the upper tail of the box plot indicates a sharp decline in GWL that might be due to excessive pumping [16]. Overall, a higher magnitude GWL trend (increasing as well as decreasing) greater than ~25–100 cm/year throughout the seasons from January to November was observed, which might be due to the effect of climate variability and/or anthropogenic activities [76–78].

Figure 2 also illustrates the magnitude of GWL variability among the selected cities. A higher magnitude of decreasing GWL trends was noticed in Allahabad (especially in August and November), followed by Jodhpur. However, in Jodhpur, a high-magnitude increasing trend was also found in some proportion of the wells throughout the seasons (January to November). A high magnitude of increasing GWL trend was detected in Rajkot, which confirms a very high rate of increasing GWL in the western arid regions of India, as reported in previous studies [16,79]. A similar magnitude of increasing and decreasing GWL was observed in Guntur from January to November. However, in May, a relatively higher decreasing GWL trend in Guntur was found, with more than 20% of wells declining at a rate greater than 10 cm/year. This might be due to a high surge in pumping during the peak of a seasonal dry spell since Guntur's economy is highly dependent on agricultural and allied activities [80]. Tiruchirappalli and Solapur exhibited similar behavior in terms of resulting slightly higher magnitude of declining GWL. Except for Rajkot, where the GWL was increasing in the majority of wells between ~2 to 20 cm/year, the overall groundwater level was found to be increasing or decreasing at ~10 cm/year in all other selected cities.

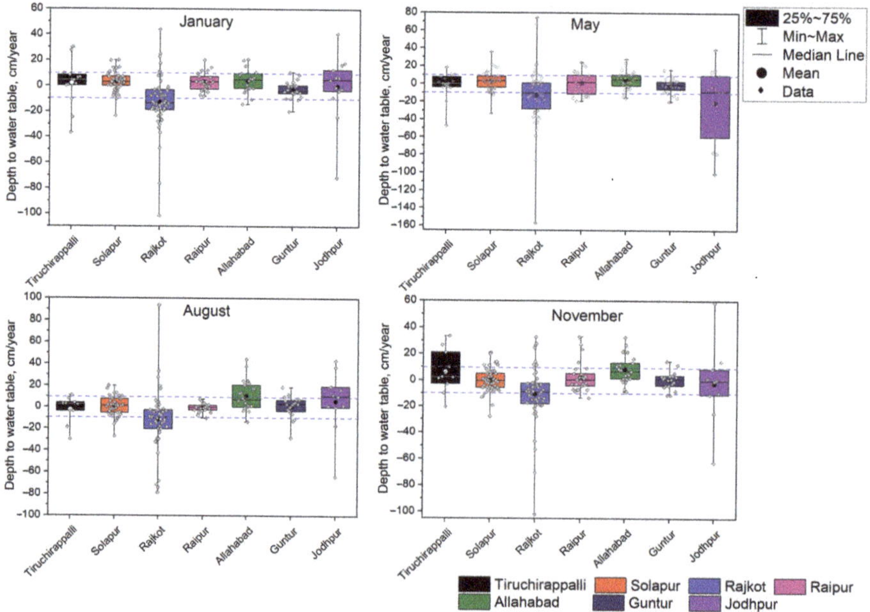

Figure 2. City-wise seasonal groundwater level (GWL) trends based on the Theil-Sen slope. A positive slope represents increasing depth to the groundwater table in cm/year, which implies decreasing GWL and vice versa. In the selected cities, GWLs were majorly varying between ±10 cm/year. Relatively, May has better GWL trends, and at the city level, Rajkot has high increasing GWL trends.

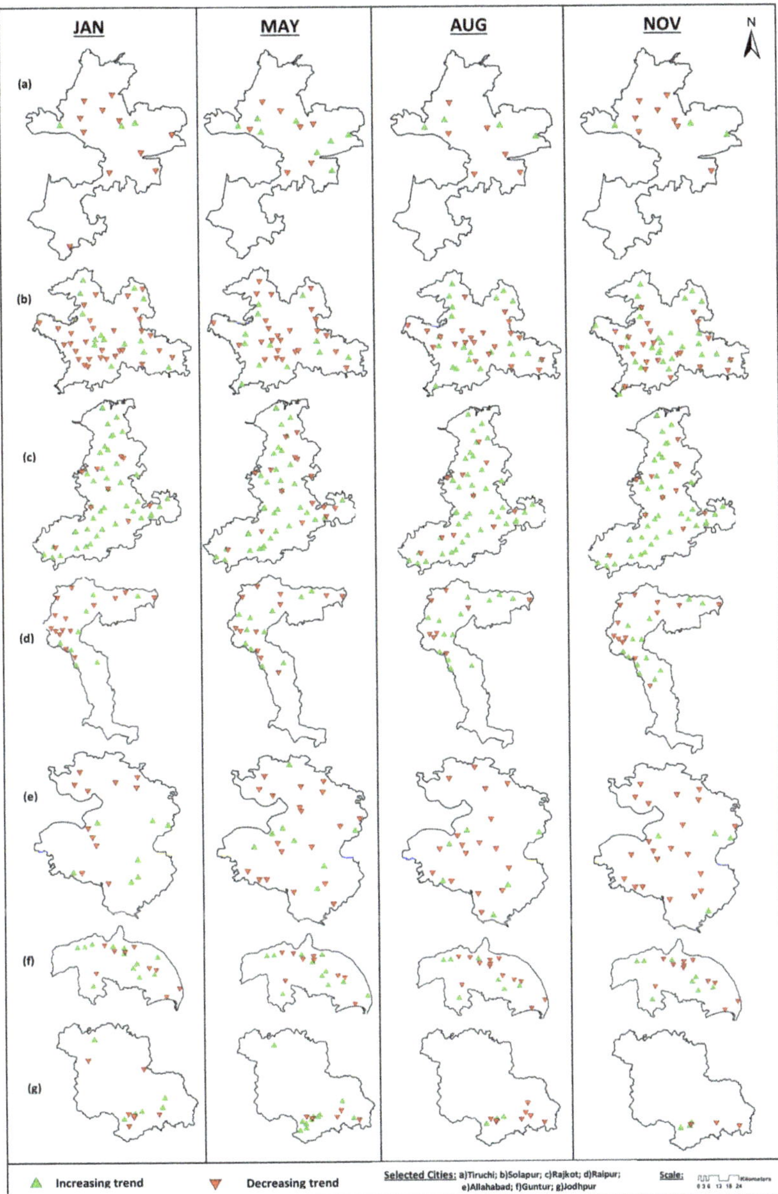

Figure 3. Spatial map depicting GWL trends for individual wells in seven cities during January, May, August, and November.

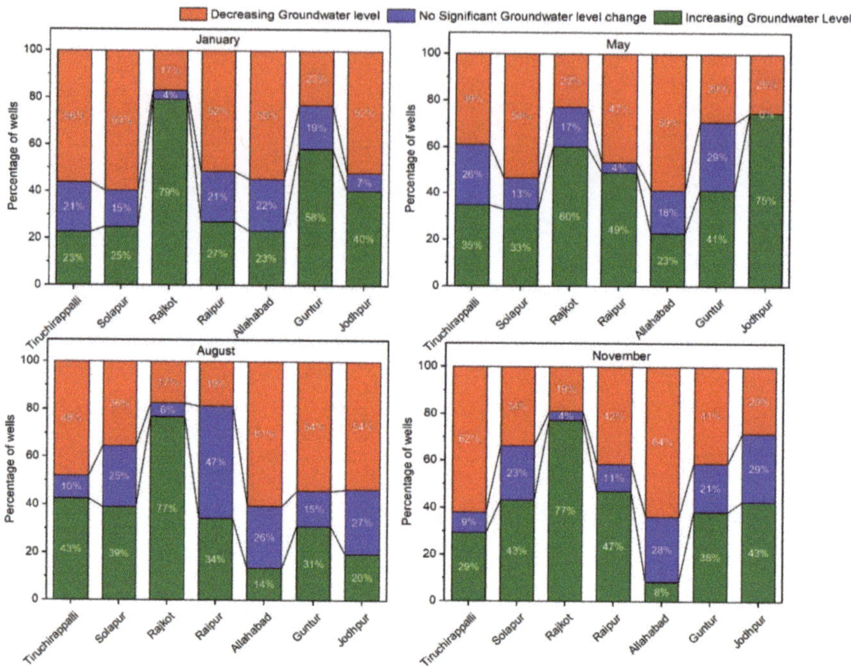

Figure 4. City-wise seasonal split-up of the wells with groundwater level trends. In majority of the selected cities, the decreasing GWL trend was found to be higher in majority of the wells throughout the season. In about ~9–47% of the wells, GWL changes occurred only between ±2 cm/year throughout the seasons. At the city level, Rajkot has the highest percentage of wells with increasing GWL and Allahabad has the highest percentage of wells with decreasing GWL with a significant trend (>2 cm/year).

There were no significant GWL changes (Figure 4) in about ~7–29% (with exception to Rajkot during January (4%), August (6%), and November (4%), and Raipur in May (4%) and August (47%)) of the wells throughout the study period (we assumed that there is no significant GWL trend if the GWL is varying at ±2 cm/year) in all these selected cities. In the selected cities, increasing and decreasing GWL trend was found in ~8–79% and ~17–64% of wells respectively, from January to November. The city-level results suggested that Rajkot has the highest (60–79%) percentage of wells with a significant increasing GWL trend (greater than 2 cm/year). In contrast, Allahabad has the highest percentage of wells (55–64%) with a significant decreasing GWL trend (greater than 2 cm/year). Raipur was determined to have an almost equal percentage of wells with significant increasing and decreasing GWL in May and November. Overall, the percentage of wells with a decreasing trend was found to be higher than the increasing trend. This reveals groundwater depletion is evident in most of the developing cities across India, which may threaten urban water security [8].

Further, analyses were conducted to find the percentage of wells having extreme increasing or decreasing GWL trends (Figure 5). Since the Theil-Sen slope estimates the median trend from a combination of slopes, it is important to note that the magnitude of the actual trend (increasing or decreasing) might be much higher than the trend estimated using the Theil-Sen slope method. Based on the thresholds of ~25 and ~50 cm/year, the percentage of wells having an extreme GWL trend during the study period was determined. In the first half of the year (i.e., January and May), the percentage of wells with increasing GWL was found to be higher than that of decreasing GWLs in the selected cities. However, at the end (November), opposite behavior was observed for which the percentage of wells having a decreasing GWL was high. This might be due to the excessive rate of pumping in November, mainly to manage the water shortage caused by the lack of rainfall during the post-monsoon season across India [74].

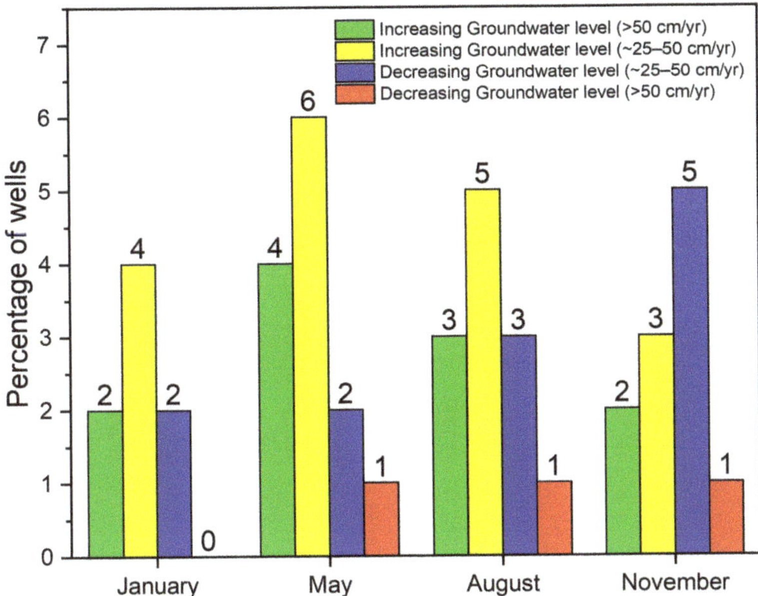

Figure 5. Percentage of wells in the seven cities with extreme (high magnitude) groundwater level trends for January, May, August, and November. About more than 5% and 2% of the wells in the selected cities have extreme increasing and decreasing GWL trends respectively, from January to November.

3.2. Non-Stationarity and Significance of Groundwater Level Trends

The significance of GWL trends was estimated using the MMK test. Based on the results (Figure 6), we found that ~40–93% of the wells in the selected cities had a significant trend during the different months of study (i.e., a monotonic trend is present, and the trend is statistically significant), with the highest MMK significance observed consistently in November (60–91% of wells showed a significant trend). Similarly, the significance of the trend at the city level was found to be in higher proportions in Solapur (79–95% of wells), followed by Tiruchirappalli (80–93% of wells, except in August). The significance of trend was found to be relatively minimal in Jodhpur and Guntur.

The combined results (Figure 6) of ADFT, KPSS, and PP tests showed that ~70–100% of the wells exhibited non-stationarity across different months (i.e., January to November). Notably, in November, about 90–100% of the wells (Figure 6) exhibited non-stationarity. At the city level, non-stationarity was observed in almost all the wells in Rajkot (98–100%) from January to November. On the other side, non-stationarity is least in Solapur, with 33% of wells (observed in August), followed by Jodhpur, which has non-stationarity in only 54% of wells during May and August. It can be understood from these results that the majority of the wells in these cities (at least more than 50%) may be significantly affected by climate variability and LULC change. The presence of such high non-stationarity would create huge uncertainty in the behavior of groundwater systems, which makes it difficult to predict future GWLs. Subsequently, this might lead to complex problems in different spheres, especially in development, public health, and irrigation, in the selected rapidly growing secondary cities [81].

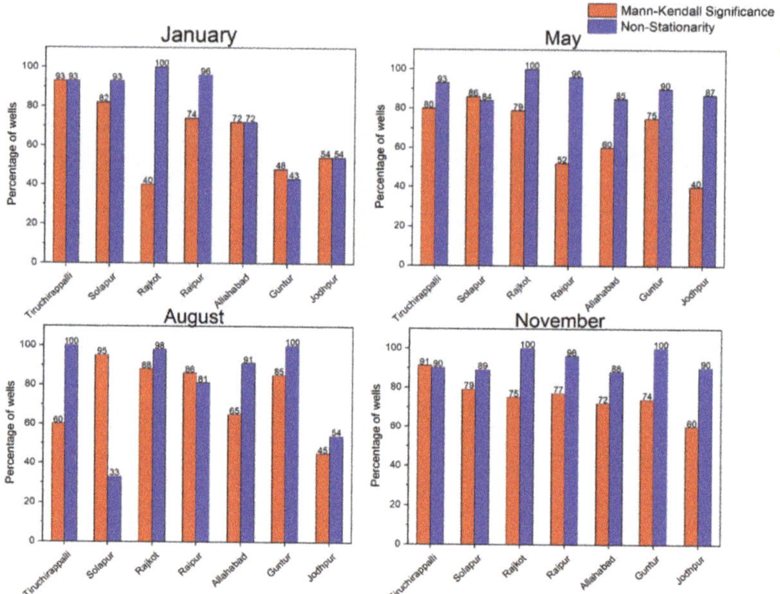

Figure 6. City-wise seasonal split-up of wells with non-stationarity and modified Mann–Kendall (MMK) trend significance in groundwater level changes.

About ~27–33% of wells in the selected cities were determined to have a decreasing GWL trend along with MMK trend significance and non-stationarity in GWL changes. At the city level, this behavior was found to be highest in Tiruchirappalli (64% of wells) and least in Rajkot (12% of wells). Notably, Jodhpur, the city with high decreasing GWL, has only ~20% of wells with this behavior. Interestingly, we determined that there was no non-stationarity as well as the MMK trend significance in wells present in the regions of high decreasing GWL trend, i.e., Rajkot and Allahabad. Although these cities (i.e., Rajkot and Allahabad) have high increasing and decreasing GWL (Figures 4 and 5, respectively), the percentage of wells having non-stationarity and MMK trend significance were relatively lower among the selected cities.

3.3. Terrestrial Water Storage Anomalies

The Theil-Sen trend in TWS anomalies (Figure 7) was estimated, and the highest declining TWS trend was observed in Jodhpur (−1.79 cm/year) (Table 2). On the contrary, TWS is highly increasing in Rajkot at 0.67 cm/year among the selected cities. While high non-stationarity was found in the GWL changes, non-stationarity in TWS anomalies was not observed in any of the selected cities. Similarly, MMK trend significance was also found to be absent in cities with a higher magnitude of increasing or decreasing TWS trends (Rajkot and Jodhpur). Unlike the GWL changes, the TWS trends indicate that the overall water storage or the TWS cycle (including surface water, soil moisture, etc., including groundwater) in the selected secondary cities have not majorly been affected by climate variability and LULC changes [51]. However, comparing the changes in TWS anomalies with GWL, a high positive correlation was observed both in the case of increasing (R = 0.85) and decreasing (R = 0.87) trends. Refer to Appendix A Figure A1 for TWS trends of all the selected cities.

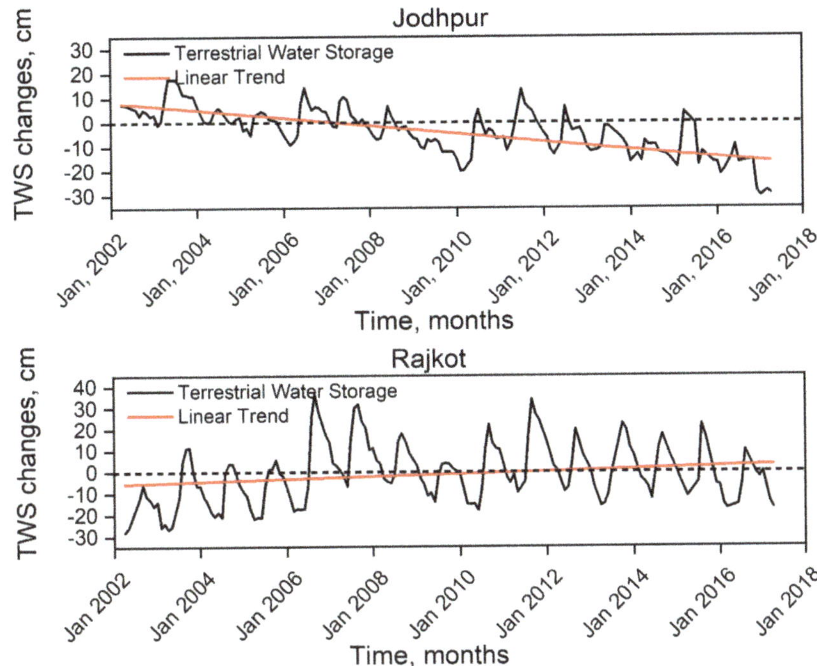

Figure 7. Terrestrial Water Storage (TWS) changes with respect to time in the cities with the high increasing (Rajkot) and decreasing (Jodhpur) GWL trends.

Table 2. Magnitude of TWS trend based on the Theil-Sen slope and significance of trend based on the MMK Test.

City	TWS Trend (cm/year)	MMK Significance for Trend	Significance Value
Tiruchirappalli	−0.51	No	0.021
Solapur	−0.67	No	0.015
Rajkot	**0.67**	**No**	**0.001**
Raipur	0.28	Yes	0.43
Jodhpur	**−1.79**	**No**	**0**
Allahabad	0.29	Yes	0.44
Guntur	−0.49	Yes	0.07

3.4. Rainfall and Groundwater Level Change Relationship

The response of GWL to rainfall was investigated through correlation analysis for the period 1996–2018. The correlation analysis on the wells having non-stationarity in the selected cities revealed that there is no strong dependency between rainfall and GWL changes. Figure 8 shows the box plot of the Correlation Coefficient (R) values for all the wells with non-stationarity. The positive R-value indicates that the GWL increases with rainfall and vice versa. The R-value varying between ±0.2 was found in ~95% of the wells for both no lag and three months lag (with the exception of Allahabad in January (under 3-month lag scenario) and Solapur and Rajkot in August (under no lag scenario)), as illustrated in Figure 8. In the 3-month lag scenario (Figure 8a), negative R values were observed in the majority of the wells, notably in Rajkot (both in January and November). A similar observation was noted for the no lag scenario (Figure 8b); however, the tail ends were shorter in the box plot in August and November. Since the majority of the wells (dug wells with depth ~10–30m) considered in this paper are shallow wells, the recharge time delay is comparatively small. Therefore, no lag scenario might be considered as a realistic scenario [71]. Although from the results of correlation analysis, no good rainfall–GWL response relationship could be drawn-out in the selected cities for January, May, and November, the R values were relatively high in Solapur, Rajkot, and Raipur in the month of August in the no lag scenario. While there is comparatively a good correlation between rainfall and GWL observed in August in the above cities, the percentage of wells with significant declining GWL trends was observed only in 20–36% of wells (Figure 4). In contrast to August, poor rainfall–GWL response was observed in January in which almost all the wells in Rajkot, Guntur, and Allahabad showed a negative correlation. January, which has a poor R-value in the no lag scenario (Figure 7b), has a relatively higher proportion of wells with decreasing GWL in Allahabad, Jodhpur, and Solapur (Figures 3 and 4), which suggests that declining GWL have a strong correlation with decreasing recharge from rainfall. The correlation analysis suggests that the rainfall and groundwater recharge relationship has been seriously affected and also ascertains the lack of seasonality in the GWL trends in the selected cities. This confirms the high variability in the rainfall events and the delay (and sometimes even failure) in the monsoon rainfall, as reported in previous studies across India [74,75,82]. The results also strongly suggest that climate variability (rainfall) has affected the GWL trends in the selected cities.

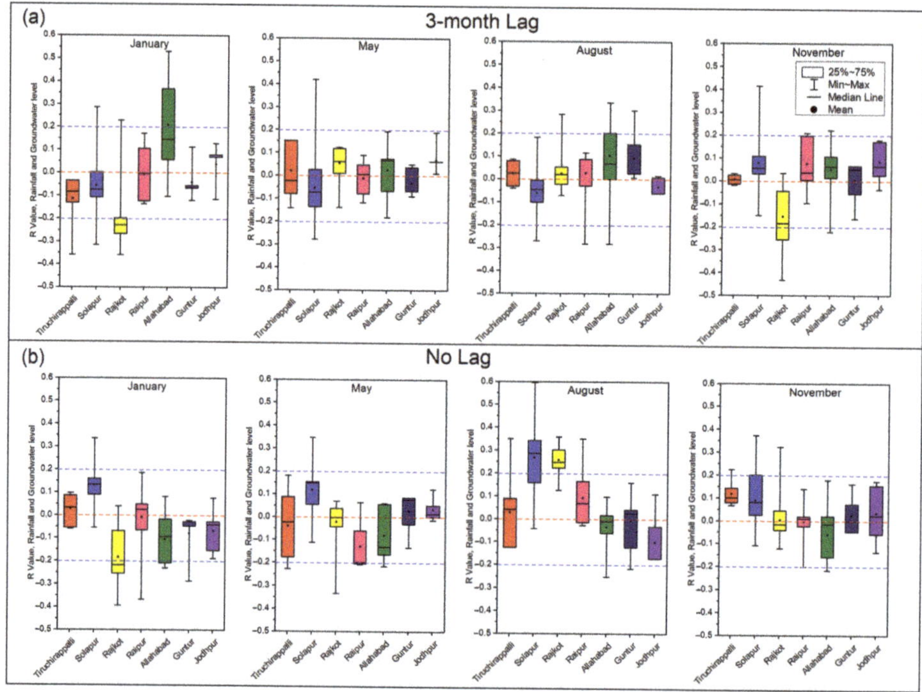

Figure 8. Correlation analysis between rainfall and groundwater level for (**a**) a 3-month recharge lag and (**b**) no lag in recharge.

3.5. Impact of Land Use/Land Cover Changes on Groundwater Level

LULC analysis (Figure 9) was performed using MODIS images collected between the years 2001 and 2018 (refer to Table 3 for detailed LULC area changes in the selected cities). Overall, the conversion of vegetation into grassland and croplands was majorly observed in all the selected cities. A relatively high growth in the urban area was seen in Raipur and Rajkot (1.6% and 2% increase, respectively), followed by Jodhpur (1.4% increase). However, in other cities, the urban area growth was less than ~0.5%, with almost no growth in the urban area of Tiruchirappalli. The higher rate of population growth (12–45% based on the census), burgeoning real-estate activities, mainly to convert agricultural fields and uncultivated land into barren land (plots) for domestic housing and industrial expansion, clearing land for road construction and highways projects, increasing the urban built area, etc. [83,84], could be considered as the main factors for the LULC changes during the study period. Notably, the high growth of the urban area in Raipur might be attributed to the city being declared as the capital of the state Chhattisgarh after its partition from Madhya Pradesh in 2001 [85]. Exponential industrial and infrastructural growth in Raipur in the past two decades has led the city to transform into a major economic hub in the region [85–87]. The LULC change trends observed in Raipur during the study could be expected in other secondary developing cities in the future, where rapid migration, infrastructural and industrial growth, etc., could lead to major LULC changes (especially the conversion of the green cover into barren land) and the sudden change might create acute stress on groundwater systems [22,87]. Unlike the metropolitans, where the growth of the urban area is very high in the outskirts of the city, the seven cities in this study show no significant expansion of urban areas outside the pre-existing urban boundary. This implies that major economic activities in the secondary Indian cities primarily take place within the existing urban boundaries, as reported in previous studies [88,89]. This subsequently indicates that most of the infrastructural developments are progressing without significantly reducing green cover in these selected cities. Within the seven cities, the highest LULC changes (Figure 9) were observed in Jodhpur, followed by Solapur and Rajkot. In Jodhpur, the cropping practice has changed, and the cropping area has increased significantly (increased by 8%) during the study period, which might be the reason for the high decreasing GWL in the region due to increased water requirement for irrigation activities [90].

On the other hand, in Rajkot, although a significant decline in the Grasslands area (green cover) occurred, the GWL trends in Rajkot are better than other cities. This trend, however, is recently observed mainly in the western arid states, which might be due to improved rainfall–GWL relationships. The positive (R = 0.14) and negative (R = −0.22) correlation coefficients were observed when cross-correlating decreasing and increasing GWL respectively, with the increase in urban areas. This strongly suggests that GWL decreases with an increase in the urban area (Table 4). These results are also in good agreement with other recent studies and suggest that LULC has a significant impact on GWL in developing cities [36,91]. Also, a very high correlation between the increase in the bare soil (decreasing green cover) area and decreasing GWL (R = 0.93) was observed. Thus, the major influence on decreasing GWL could be attributed to the conversion of the green cover into barren land for various anthropogenic activities such as infrastructural development, real-estate, etc. It may be inferred from these results that the irrigation and agricultural water requirements, although expected to have a very high impact on decreasing GWL in the selected developing cites (since most of the selected cities are still having major agricultural activities outside the urban boundary), are not the major driver; however, the conversion of the green cover (majorly the cropping area) into barren land has more influence on decreasing GWL. LULC change analyses also reveal that an increase in the urban area has a significant impact on decreasing GWL. However, the extent of the impact is comparatively lower (R = 0.14), which could be due to the proper functioning of the public water supply schemes and lower domestic water demands. However, as the urban area increases in the future, the stress on GWL might also increase in the selected cities, as previously observed in other already developed Indian cities [92,93].

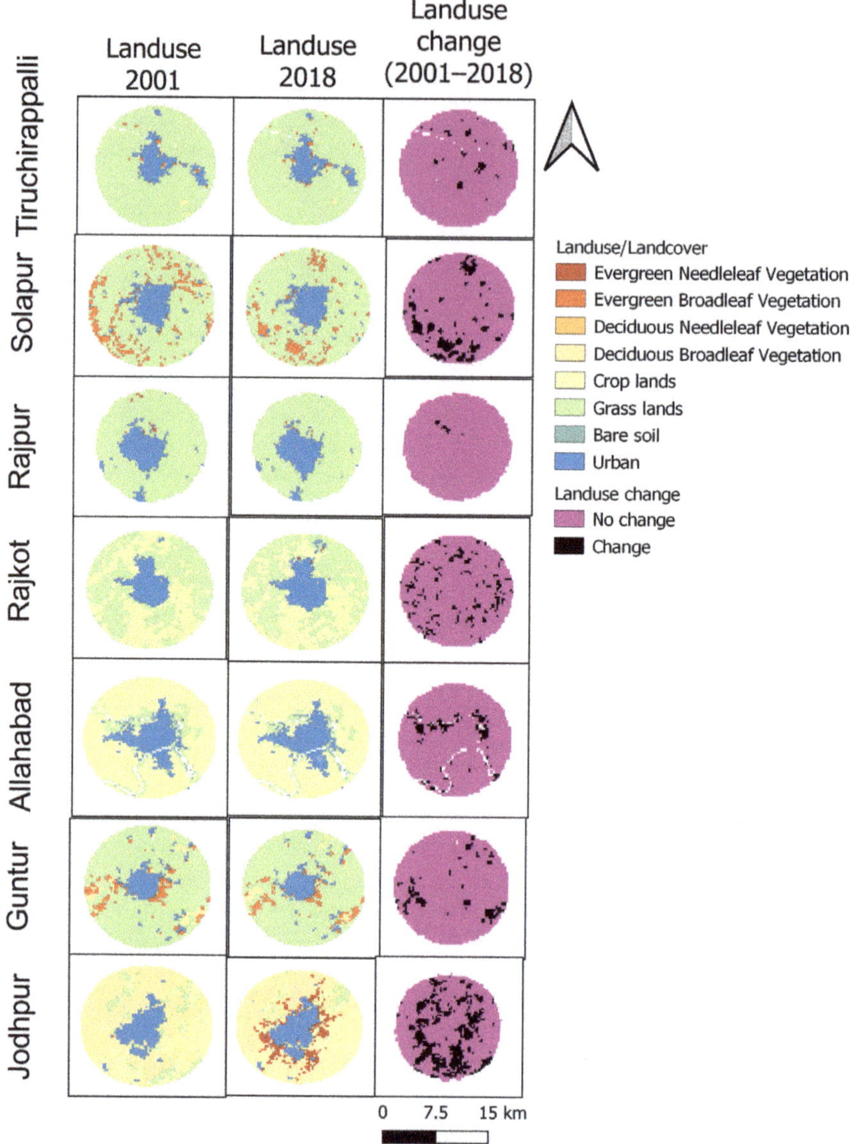

Figure 9. Land Use/Land Cover (LULC) changes in the selected cities.

Table 3. Land use/land cover changes between 2001 and 2018 in the selected cities.

Cities	Land Use/Land Cover	2001		2018		Land-Use Change (2001–2018)	
		Area (km^2)	Area (%)	Area (km^2)	Area (%)	Area (km^2)	Area (%)
Rajkot	Evergreen needle leaf vegetation	2.7	0.4	0.0	0.0	−2.7	−0.4
	Grass lands	585.4	84.0	582.2	82.3	−3.3	−1.7
	Bare soil	0.0	0.0	0.9	0.1	0.9	0.1
	Urban	**108.7**	**15.6**	**124.1**	**17.6**	**15.4**	**2.0**
	Crop lands	254.6	37.6	256.4	37.9	1.9	0.3
Raipur	Grass lands	337.0	49.8	324.0	47.9	−13.0	−1.9
	Urban	**85.2**	**12.6**	**96.3**	**14.2**	**11.1**	**1.6**
	Evergreen broad leaf vegetation	71.2	10.3	47.8	6.9	−23.4	−3.4
Solapur	Grass lands	537.6	77.3	562.9	81.0	25.3	3.7
	Urban	86.2	12.4	84.3	12.1	−1.9	−0.3
	Evergreen needle leaf vegetation	0.0	0.0	59.0	8.6	59.0	8.6
	Evergreen broad leaf vegetation	0.0	0.0	2.8	0.4	2.8	0.4
Jodhpur	Deciduous broad leaf vegetation	365.2	52.9	260.4	37.7	−104.9	−15.2
	Crop lands	**176.1**	**25.5**	**233.2**	**33.8**	**57.1**	**8.3**
	Grass lands	60.9	8.8	36.5	5.3	−24.4	−3.5
	Urban	**88.0**	**12.8**	**98.3**	**14.2**	**10.3**	**1.4**
Tiruchirappalli	Evergreen needle leaf vegetation	0.3	0.0	0.3	0.0	0.0	0.0
	Evergreen broad leaf vegetation	6.8	0.9	9.5	1.3	2.8	0.4
	Deciduous broad leaf vegetation	3.5	0.5	3.5	0.5	0.0	0.0
	Grass lands	651.8	87.4	649.0	87.0	−2.8	−0.4
	Urban	83.8	11.2	83.8	11.2	0.0	0.0
Guntur	Evergreen Broad leaf Vegetation	51.0	6.6	47.5	6.2	−3.5	−0.5
	Deciduous Broad leaf Vegetation	29.3	3.8	19.3	2.5	−10.0	−1.3
	Grass lands	611.0	79.7	623.8	81.3	12.8	1.7
	Urban	75.8	9.9	76.5	10.0	0.8	0.1
Allahabad	Crop lands	549.5	69.7	573.3	72.7	23.8	3.0
	Grass lands	98.8	12.5	72.3	9.2	−26.5	−3.4
	Barren soil	13.3	1.7	12.0	1.5	−1.3	−0.2
	Urban	127.3	16.1	131.3	16.6	4.0	0.5

Table 4. Correlation coefficient (R) values for change in the area of different LULC classes with the groundwater level in the selected cities.

LULC Class	R-Value: Wells with Increasing GWL Trends	R-Value: Wells with Decreasing GWL Trends
Evergreen Needle leaf Vegetation	0.32	−0.59
Evergreen Broadleaf Vegetation	−0.20	0.05
Deciduous Needle leaf Vegetation	-	-
Deciduous Broadleaf Vegetation	−0.27	0.72
Cropland	0.19	−0.69
Grass land	−0.21	0.41
Bare Soil	**−0.97**	**0.93**
Urban	**−0.22**	**0.14**

4. Conclusions

This paper focused on understanding the GWL trends in the secondary developing cities in India. Seven cities were selected for the analysis based on various aspects, including the socio-economic and population growth trajectories. The key findings were derived based on statistical and LULC change analysis performed on the GWL data. Overall, it was found that the GWL varies between ±10 cm/year, with the trend to be significant (MMK test) in the majority of the wells. Non-stationarity was observed in the majority of the wells throughout the seasons, which suggests GWL has been affected by both climate variability (in monsoon rainfall) and LULC change. However, wells having a high decreasing GWL trend mostly did not exhibit non-stationarity in GWL changes. This indicates that the declining GWL of higher magnitudes has no substantial variation, possibly due to irreversible aquifer depletion caused by a high rate of pumping over a long time, where the effects of climate variability and LULC changes might be negligible. Less interdependence between seasonality (based on monsoon) and GWL trends was observed, which might have been caused by the impact of climate variability on groundwater recharge. The cause of non-stationarity in GWL was analyzed by linking climate variability (rainfall) and LULC with the GWL changes. Correlation analysis between GWL and rainfall indicates that the rainfall–groundwater level relationship has been seriously affected across all the selected cities, plausibly due to the failure of monsoon rainfall. Based on LULC change analysis, a high correlation between decreasing green cover (increasing barren land) (R = 0.93) and decreasing GWL indicates a high anthropogenic activity. Although agriculture-based activities are still actively practiced in secondary developing cities, its impact on GWL was surprisingly relatively low. The change in the urban area was majorly within the pre-existing urban boundary, and the domestic water requirement presently might not contribute much to the decreasing GWL. However, with more urbanization, this might change in the future. The major findings of this study can very well be adopted in other secondary developing cities with similar socio-economic growth patterns, especially in developing countries.

Instead of using the derived GWLs from satellite products, which are considered to contain more error and uncertainty, this paper used point scale well observations to assess the trends and non-stationarity in GWL changes. However, the bias in the trend and non-stationarity estimates are still possible due to the limitations in the data length and the number of available observation wells. Therefore, future research works, including the more observed data to conduct a similar study, are recommended.

Author Contributions: Conceptualization, K.S.K. and B.-S.S.; methodology, K.S.K., J.H. and S.M.; software, K.S.K., S.M., M.M.D.M. and A.M.; validation, I.I. and K.S.K.; formal analysis, K.S.K. and S.M.; investigation, K.S.K., J.H., S.M.P., B.-S.S., S.M. and I.I.; data curation, A.M.; Writing, K.S.K., I.I. and B.-S.S.; Review, K.S.K., J.H., S.M.P., B.-S.S., M.M.D.M. and I.I.; visualization, K.S.K. and I.I.; supervision, K.S.K. All authors have read and agreed to the published version of the manuscript.

Funding: This research received no external funding.

Acknowledgments: The authors wish to thank the Editor and four anonymous reviewers for their constructive comments and suggestions, which helped to improve the complete structure of the paper. This research work was supported by the Early Career Research Award from SERB, Government of India (File NO.ECR/2018/002136), received by the corresponding author.

Conflicts of Interest: The authors declare no conflict of interest.

Appendix A

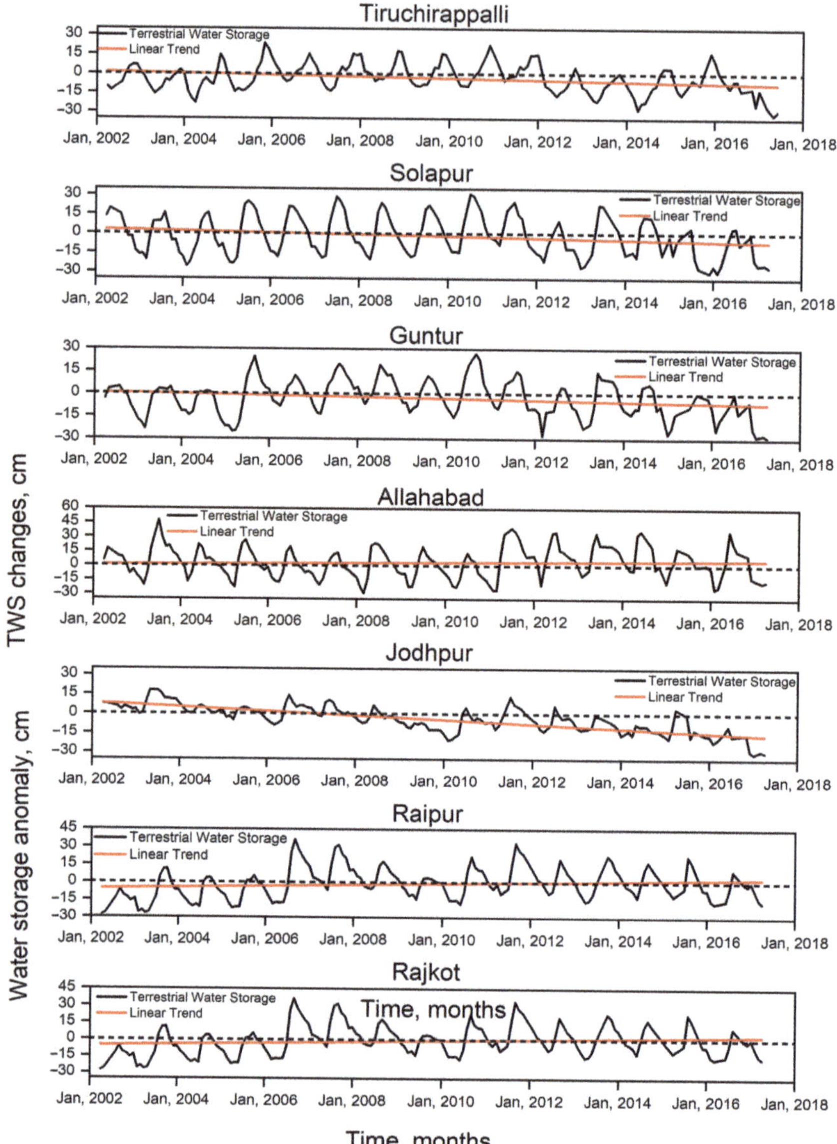

Figure A1. Terrestrial Water Storage (TWS) changes w.r.t time in the selected cities.

References

1. Margat, J.; Gun, J. *Groundwater around the World*, 1st ed.; CRC Press: London, UK, 2013. [CrossRef]
2. Mullen, K. Groundwater Information on Earth's Water. Available online: https://www.ngwa.org/what-is-groundwater/About-groundwater/information-on-earths-water (accessed on 26 October 2020).
3. Zektser, I.S.; Everett, L.G. Groundwater Resources of the World and Their Use. UNESCO Digital Library. Available online: https://unesdoc.unesco.org/ark:/48223/pf0000134433 (accessed on 14 June 2020).

4. Siebert, S.; Burke, J.; Faures, J.; Frenken, K.; Hoogeveen, J.; Döll, P.; Portmann, F. Groundwater use for irrigation—A global inventory. *Hydrol. Earth Syst. Sci.* **2010**, *14*, 1863–1880. [CrossRef]

5. Vörösmarty, C.; McIntyre, P.; Gessner, M.; Dudgeon, D.; Prusevich, A.; Green, P. Global threats to human water security and river biodiversity. *Nature* **2010**, *467*, 555–561. [CrossRef]

6. 68% of the World Population Projected to Live in Urban Areas by 2050, Says U.N. Available online: https://www.un.org/development/desa/en/news/population/2018-revision-of-world-urbanization-prospects.html (accessed on 26 October 2020).

7. Elfie, S.; Denise, P.; Eric, D. The future of India's urbanization. *Futures* **2014**, *56*, 43–52. [CrossRef]

8. Foster, S.; Hirata, R. Groundwater use for urban development: Enhancing benefits and reducing risks. *Water Front.* **2011**, *2011*, 21–29.

9. Howard, K.; Hirata, R.; Warner, K.; Gogu, R.; Nkhuwa, D. Resilient Cities & Groundwater. In *IAH Strategic Overview Series*; Stephen, F., Gillian, T., Eds.; International Association of Hydrogeologists: London, UK, 2015. Available online: https://www.iges.or.jp/en/pub/resilient-cities-groundwater/en (accessed on 26 October 2020).

10. The United Nations World Water Development Report 2015: Water for a Sustainable World. Available online: https://unesdoc.unesco.org/ark:/48223/pf0000231823 (accessed on 13 November 2020).

11. Donald, R.I.; Katherine, W.; Julie, P.; Martina, F.; Christof, S. Water on an urban planet: Urbanization and the reach of urban water infrastructure. *Glob. Environ. Chang.* **2014**, *27*, 96–105. [CrossRef]

12. Bricker, S.H.; Banks, V.J.; Galik, G. Accounting for groundwater in future city visions. *Land Use Policy* **2017**, *69*, 618–630. [CrossRef]

13. Foster, S.D.; Ricardo, H.; Ken, W.F.H. Groundwater use in developing cities: Policy issues arising from current trends. *Hydrogeol. J.* **2010**, *19*, 271–274. [CrossRef]

14. Tiwari, V.; Wahr, J.; Swenson, S. Dwindling groundwater resources in northern India, from satellite gravity observations. *Geophys. Res. Lett.* **2009**, *36*. [CrossRef]

15. Chatterjee, R.; Purohit, R. Estimation of Replenishable Groundwater Resources of India and Their Status of Utilization. *Curr. Sci.* **2009**, *96*, 1581–1591.

16. Asoka, A.; Gleeson, T.; Wada, Y.; Mishra, V. Relative contribution of monsoon precipitation and pumping to changes in groundwater storage in India. *Nat. Geosci.* **2017**, *10*, 109–117. [CrossRef]

17. Rodell, M.; Velicogna, I.; Famiglietti, J. Satellite-based estimates of groundwater depletion in India. *Nature* **2009**, *460*, 999–1002. [CrossRef]

18. Hora, T.; Srinivasan, V.; Basu, N. The Groundwater Recovery Paradox in South India. *Geophys. Res. Lett.* **2019**, *46*, 9602–9611. [CrossRef]

19. Gana, K. Uncovering the Groundwater Crisis in South India. Available online: https://india.mongabay.com/2019/09/uncovering-the-groundwater-crisis-in-south-india (accessed on 26 October 2020).

20. Singh, R.B. Impact of land-use change on groundwater in the Punjab-Haryana plains, India. In Proceedings of the Impact of Human Activity on Groundwater Dynamics, Maastricht, The Netharlands, 18 July 2001; IAHS-AISH Publication: Wallingford, UK, 2001; pp. 117–122.

21. Samanpreet, B.; Rajan, A.; Mandeep, B. Groundwater Depletion in Punjab, India. In *Encyclopedia of Soil Science*, 3rd ed.; Tylor & Francis: Abingdon-on-Thames, UK, 2017; pp. 1–5. [CrossRef]

22. Foster, S. Global Policy Overview of Groundwater in Urban Development—A Tale of 10 Cities. *Water* **2020**, *12*, 456. [CrossRef]

23. Ken, W.H.F. *Urban Groundwater: Meeting the Challenge*; IAH Selected Paper Series 8; CRC Press: Boca Raton, FL, USA, 2007; ISBN 9780415407458.

24. NITI. Aayog Composite Water Management Index 2.0. Available online: https://niti.gov.in/sites/default/files/2019-08/CWMI-2.0-latest.pdf (accessed on 26 October 2020).

25. Chennai Water Crisis: City's Reservoirs Run Dry. Available online: https://www.bbc.com/news/world-asia-india-48672330 (accessed on 26 October 2020).

26. Service Level Benchmarks. Available online: http://mohua.gov.in/cms/Service-Level-Benchmarks.php (accessed on 26 October 2020).

27. Gontoju, S.S.; Alam, M.F.; Sikka, A. Chennai Water Crisis: A Wake-Up Call for Indian Cities. Available online: https://www.downtoearth.org.in/blog/water/chennai-water-crisis-a-wake-up-call-for-indian-cities-66024 (accessed on 18 October 2020).

28. Aspinwall, N. Chennai's 'Man-Made' Water Crisis. Available online: https://thediplomat.com/2019/08/chennais-man-made-water-crisis (accessed on 25 October 2020).

29. World Water Day: India is 3rd Largest Groundwater Exporter, but 21 Cities are Running out of Water by 2020. Available online: https://www.indiatoday.in/science/story/world-water-day-2019-water-crisis-india-1483777-2019-03-22 (accessed on 26 October 2020).

30. Singh, S.; Sharma, V. Urban Droughts in India: Case Study of Delhi. *Disaster Risk Reduct.* **2018**, 155–167. [CrossRef]

31. Sinan, M.; Mishra, V. Urban Drought in India under Observed and Projected Future Climate. In Proceedings of the AGU Fall Meeting, Washington, DC, USA, 10–14 December 2018.

32. Alipour, A.; Hashemi, S.; Shokri, S.; Moravej, M. Spatio-temporal analysis of groundwater level in an arid area. *Int. J. Water* **2018**, *12*, 66. [CrossRef]

33. Jessica, H.; Katherine, C.; George, K. Tools to Estimate Groundwater Levels in the Presence of Changes of Precipitation and Pumping. *J. Water Resour. Prot.* **2016**, *8*, 12. [CrossRef]

34. Gurdak, J.J.; Kuss, A.M. Teleconnections in the Principal Aquifers to Non-Stationarity of ENSO, NAO, PDO, and AMO. In Proceedings of the AGU Fall Meeting, Washington, DC, USA, 10–14 December 2012.

35. Graf, R. Reference statistics for the structure of measurement series of groundwater levels (Wielkopolska Lowland, western Poland). *Hydrol. Sci. J.* **2015**, *60*, 1587–1606. [CrossRef]

36. Kayet, N.; Chakrabarty, A.; Pathak, K.; Sahoo, S.; Mandal, S.; Fatema, S. Spatiotemporal LULC change impacts on groundwater table in Jhargram, West Bengal, India. *Sustain. Water Resour. Manag.* **2018**, *5*, 1189–1200. [CrossRef]

37. Roy, A.; Inamdar, A. Multi-temporal Land Use Land Cover (LULC) change analysis of a dry semi-arid river basin in western India following a robust multi-sensor satellite image calibration strategy. *Heliyon* **2019**, *5*, e01478. [CrossRef]

38. Pranab, K.S. Estimates of the Regression Coefficient Based on Kendall's Tau. *J. Am. Stat. Assoc.* **1968**, *63*, 1379–1389. [CrossRef]

39. Hamed, K.; Rao, A.R. A modified Mann-Kendall trend test for autocorrelated data. *J. Hydrol.* **1998**, *204*, 182–196. [CrossRef]

40. Dickey, D.A.; Fuller, W.A. Distribution of the estimation of autoregressive time series with a unit root. *J. Am. Stat. Assoc.* **1979**, *74*, 427–431.

41. Phillips, P. Testing for a unit root in time series regression. *Biometrika* **1988**, *75*, 335–346. [CrossRef]

42. Kwiatkowski, D.; Phillips, P.C.B.; Schmidt, P.; Shin, Y. Testing the null hypothesis of stationarity against the alternative of a unit root. *J. Econom.* **1992**, *54*, 159–178. [CrossRef]

43. Fathian, F.; Saeed, M.; Ercan, K. Identification of trends in hydrological and climatic variables in Urmia Lake Basin, Iran. *Theor. Appl. Climatol.* **2014**, *119*, 443–464. [CrossRef]

44. Navas, R.; Alonso, J.; Gorgoglione, A.; Vervoort, R.W. Identifying climate and human impact trends in streamflow: A case study in Uruguay. *Water* **2019**, *11*, 1433. [CrossRef]

45. Census of India. Available online: https://censusindia.gov.in (accessed on 26 October 2020).

46. Mapping India's GDP. Available online: https://www.idfcinstitute.org/projects/transitions/city-gdp/ (accessed on 20 October 2020).

47. Smart Cities Mission. Available online: http://smartcities.gov.in/content/ (accessed on 26 October 2020).

48. Water Resource Information System, India-WRIS. Available online: https://indiawris.gov.in/wris/#/groundWater (accessed on 26 October 2020).

49. Central Ground Water Board. Available online: http://cgwb.gov.in/ (accessed on 26 October 2020).

50. Grubbs, F.E. Procedures for Detecting Outlying Observations in Samples. *Technimetrics* **1916**, *11*, 1–21. [CrossRef]

51. Monthly Mass Grids GRACE Tellus Data. Available online: https://grace.jpl.nasa.gov/data/get-data/monthly-mass-grids-land (accessed on 25 October 2020).

52. Pai, D.S.; Latha, S.; Rajeevan, M. Development of a new high spatial resolution (0.25° × 0.25°) Long period (1901–2010) daily gridded rainfall data set over India and its comparison with existing data sets over the region. *Mausam* **2014**, *65*, 1–18.

53. Friedl, M.; Sulla-Menashe, D. *MCD12Q1 MODIS/Terra+Aqua Land Cover Type Yearly L3 Global 500m SIN Grid V006*; NASA EOSDIS Land Processes DAAC: Sioux Falls, SD, USA, 2015. [CrossRef]

54. Yue, S.; Pilon, P.; Cavadias, G. Power of the Mann–Kendall and Spearman's rho tests for detecting monotonic trends in hydrological series. *J. Hydrol.* **2002**, *259*, 254–271. [CrossRef]

55. Ali, R.; Kuriqi, A.; Abubaker, S.; Kisi, O. Long-Term Trends and Seasonality Detection of the Observed Flow in Yangtze River Using Mann-Kendall and Sen's Innovative Trend Method. *Water* **2019**, *11*, 1855. [CrossRef]

56. Mann, H.B. Nonparametric tests against trend. *Econometrica* **1945**, *13*, 245–259. [CrossRef]

57. Kendall, M.G. *Rank Correlation Methods*, 4th ed.; Charles, G., Ed.; Griffin: Brighton, London, UK, 2020; ISBN 0852641990.

58. Chen, J.; Famiglietti, J.; Scanlon, B.; Rodell, M. Erratum to: Groundwater Storage Changes: Present Status from GRACE Observations. *Surv. Geophys.* **2016**, *37*, 701. [CrossRef]

59. Klees, R.; Zapreeva, E.; Winsemius, H.; Savenije, H. The bias in GRACE estimates of continental water storage variations. *Hydrol. Earth Syst. Sci.* **2007**, *11*, 1227–1241. [CrossRef]

60. Bayazit, M. Nonstationarity of Hydrological Records and Recent Trends in Trend Analysis: A State-of-the-art Review. *Environ. Process.* **2015**, *2*, 527–542. [CrossRef]

61. Augmented Dickey-Fuller Test. Available online: https://www.mathworks.com/help/econ/adftest.html (accessed on 27 October 2020).

62. Diebold, F.X. Unit Root Tests Are Useful for Selecting Forecasting Models. *NBER Working Papers J. Bus. Econ. Stat.* **2000**, *18*, 265–273.

63. Phillips-Perron Test for One Unit Root—MATLAB Pptest. Available online: https://www.mathworks.com/help/econ/pptest.html (accessed on 27 October 2020).

64. KPSS Test for Stationarity—MATLAB Kpsstest. Available online: https://www.mathworks.com/help/econ/kpsstest.html (accessed on 24 October 2020).

65. Arltová, M.; Fedorová, D. Selection of Unit Root Test on the Basis of Length of the Time Series and Value of AR(1) Parameter. *Statistika* **2016**, *96*, 47–64.

66. Zivot, E.; Wang, J. Unit Roots Tests. In *Modeling Financial Time Series with S-PLUS®*; Springer: New York, NY, USA, 2016; pp. 111–139. [CrossRef]

67. Wu, J.; Zhang, R.; Yang, J. Analysis of rainfall–recharge relationships. *J. Hydrol.* **1996**, *177*, 143–160. [CrossRef]

68. Cai, Z.; Ofterdinger, U. Analysis of groundwater-level response to rainfall and estimation of annual recharge in fractured hard rock aquifers, NW Ireland. *J. Hydrol.* **2016**, *535*, 71–84. [CrossRef]

69. Bharathkumar, L.; Mohammed-Aslam, M.A. Long-Term Trend Analysis of Water-Level Response to Rainfall of Gulbarga Watershed, Karnataka, India, In Basaltic Terrain: Hydrogeological Environmental Appraisal in Arid Region. *Appl. Water Sci.* **2018**, *8*. [CrossRef]

70. Şen, Z. Groundwater Recharge Level Estimation from Rainfall Record Probability Match Methodology. *Earth Syst. Environ.* **2019**, *3*, 603–612. [CrossRef]

71. Hocking, M.; Kelly, B. Groundwater recharge and time lag measurement through Vertosols using impulse response functions. *J. Hydrol.* **2016**, *535*, 22–35. [CrossRef]

72. Ross, R.; Krishnamurti, T.; Pattnaik, S.; Pai, D. Decadal surface temperature trends in India based on a new high-resolution data set. *Sci. Rep.* **2018**, *8*. [CrossRef] [PubMed]

73. Ghatak, D.; Zaitchik, B.; Hain, C.; Anderson, M. The role of local heating in the 2015 Indian Heat Wave. *Sci. Rep.* **2017**, *7*, 1–8. [CrossRef] [PubMed]

74. Dash, S.; Kulkarni, M.; Mohanty, U.; Prasad, K. Changes in the characteristics of rain events in India. *J. Geophys. Res.* **2009**, *114*, 114. [CrossRef]

75. Ghosh, S.; Vittal, H.; Sharma, T.; Karmakar, S.; Kasiviswanathan, K. Indian Summer Monsoon Rainfall: Implications of Contrasting Trends in the Spatial Variability of Means and Extremes. *PLoS ONE* **2016**, *11*, e0158670. [CrossRef]

76. Dunne, D. Climate Change's Impact on Groundwater Could Leave 'Environmental Time Bomb'. Available online: https://www.carbonbrief.org/climate-change-impact-groundwater-environmental-timebomb (accessed on 26 October 2020).

77. Cuthbert, M.; Gleeson, T.; Moosdorf, N.; Befus, K.; Schneider, A.; Hartmann, J.; Lehner, B. Global patterns and dynamics of climate–groundwater interactions. *Nat. Clim. Chang.* **2019**, *9*, 137–141. [CrossRef]

78. Bhanja, S.; Mukherjee, A.; Rodell, M.; Velicogna, I.; Pangaluru, K.; Famiglietti, J. Regional storage changes in the Indian subcontinent: The role of anthropogenic activities. In Proceedings of the AGU Fall Meeting, San Francisco, CA, USA, 15–19 December 2014. Available online: https://ui.adsabs.harvard.edu/abs/2014AGUFMGC21B0533B/abstract (accessed on 22 October 2020).

79. Kumar, M.D.; Perry, C.J. What can explain groundwater rejuvenation in Gujarat in recent years? *Int. J. Water Resour. Dev.* **2014**, *35*, 891–906. [CrossRef]

80. Economy of Guntur District. Economic Development Board Andhra Pradesh—Infrastructure. Available online: http://apedb.gov.in/infrastrctr.html (accessed on 25 October 2020).

81. Groundwater Resilience to Human Development and Climate Change in South Asia. Available online: https://globalwaterforum.org/2013/08/19/groundwater-resilience-to-human-development-and-climate-change-in-south-asia (accessed on 25 October 2020).

82. Loo, Y.; Billa, L.; Singh, A. Effect of climate change on seasonal monsoon in Asia and its impact on the variability of monsoon rainfall in Southeast Asia. *Geosci. Front.* **2015**, *6*, 817–823. [CrossRef]

83. Gogoi, P.; Vinoj, V.; Swain, D.; Roberts, G.; Dash, J.; Tripathy, S. Land use and land cover change effect on surface temperature over Eastern India. *Sci. Rep.* **2019**, *9*, 1–10. [CrossRef]

84. Madhavi, J.; Dorje, D.; Rajender, M.; Dimri, A.P. Monitoring Land use change and its drivers in Delhi, India using multi-temporal satellite data. *Model. Earth Syst. Environ.* **2016**, *2*, 19. [CrossRef]

85. Raipur City, the Capital of Chhattisgarh. Available online: https://www.indyatour.com/india/chhattisgarh/tour-spots/raipur-capital-city-chhattisgarh.php (accessed on 20 October 2020).

86. How Naya Raipur Is Emerging as One of the Well-Planned Smart Cities in India. Available online: https://www.thebetterindia.com/95266/naya-raipur-smart-city-chattisgarh (accessed on 24 October 2020).

87. Naya Raipur—India's 4th Planned Capital City Is Developing in Full Swing. Available online: https://worldarchitecture.org/articles/cvegh/naya_raipur__india_s_4th_planned_capital_city_is_developing_in_full_swing.html (accessed on 26 October 2020).

88. Tewari, M.; Godfrey, N. Better Cities, Better Growth: India's Urban Opportunity. Available online: http://newclimateeconomy.report/workingpapers (accessed on 20 October 2020).

89. Vishwanath, T.; Lall, S.V.; Dowall, D.; Lozano-Gracia, N. Urbanization beyond Municipal Boundaries: Nurturing Metropolitan Economies and Connecting Peri-Urban Area. Available online: http://documents.worldbank.org/curated/en/373731468268485378/Urbanization-beyond-municipal-boundaries-nurturing-metropolitan-economies-and-connecting-peri-urban-areas-in-India (accessed on 25 October 2020).

90. Chinnasamy, P.; Maheshwari, B.; Prathapar, S. Understanding Groundwater Storage Changes and Recharge in Rajasthan, India through Remote Sensing. *Water* **2015**, *7*, 5547–5565. [CrossRef]

91. Prabhakar, A.; Tiwari, H. Land use and land cover effect on groundwater storage. *Model. Earth Syst. Environ.* **2015**, *1*, 1–10. [CrossRef]

92. Arfanuzzaman, M.; Atiq Rahman, A. Sustainable water demand management in the face of rapid urbanization and ground water depletion for social–ecological resilience building. *Glob. Ecol. Conserv.* **2017**, *10*, 9–22. [CrossRef]

93. Le Vo, P. Urbanization and water management in Ho Chi Minh City, Vietnam-issues, challenges and perspectives. *Geojournal* **2007**, *70*, 75–89. [CrossRef]

MDPI

St. Alban-Anlage 66

4052 Basel

Switzerland

Tel. +41 61 683 77 34

Fax +41 61 302 89 18

www.mdpi.com

Water Editorial Office

E-mail: water@mdpi.com

www.mdpi.com/journal/water

9 783036 508108